Erwin Dee Kord (Ed.)

Seneca, Oregon

Erwin Dee Kord (Ed.)

Seneca, Oregon

Grant County

Solv

Contents

Articles

Seneca,_Oregon	1
Grant_County,_Oregon	4
Oregon	12
Blue_Mountains_(Oregon)	36
Canyon_City,_Oregon	37
U.S._Route_395_in_Oregon	40
Malheur_National_Forest	42
Oregon_and_Northwestern_Railroad	44
Pinus ponderosa	46
Hines,_Oregon	51
Silvies_River	54
Great_Basin	56

References

Article Sources and Contributors	61
Image Sources, Licenses and Contributors	63

Seneca,_Oregon

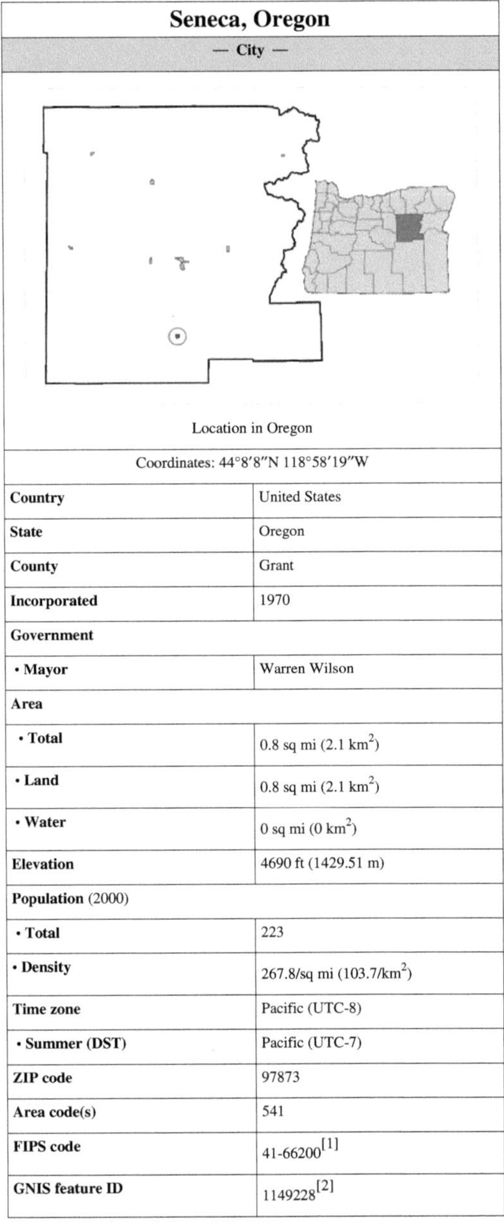

<table>
<tr><td colspan="2" align="center">Seneca, Oregon</td></tr>
<tr><td colspan="2" align="center">— City —</td></tr>
<tr><td colspan="2" align="center">Location in Oregon</td></tr>
<tr><td colspan="2" align="center">Coordinates: 44°8′8″N 118°58′19″W</td></tr>
<tr><td>Country</td><td>United States</td></tr>
<tr><td>State</td><td>Oregon</td></tr>
<tr><td>County</td><td>Grant</td></tr>
<tr><td>Incorporated</td><td>1970</td></tr>
<tr><td colspan="2">Government</td></tr>
<tr><td> • Mayor</td><td>Warren Wilson</td></tr>
<tr><td colspan="2">Area</td></tr>
<tr><td> • Total</td><td>0.8 sq mi (2.1 km^2)</td></tr>
<tr><td> • Land</td><td>0.8 sq mi (2.1 km^2)</td></tr>
<tr><td> • Water</td><td>0 sq mi (0 km^2)</td></tr>
<tr><td>Elevation</td><td>4690 ft (1429.51 m)</td></tr>
<tr><td colspan="2">Population (2000)</td></tr>
<tr><td> • Total</td><td>223</td></tr>
<tr><td> • Density</td><td>267.8/sq mi (103.7/km^2)</td></tr>
<tr><td>Time zone</td><td>Pacific (UTC-8)</td></tr>
<tr><td> • Summer (DST)</td><td>Pacific (UTC-7)</td></tr>
<tr><td>ZIP code</td><td>97873</td></tr>
<tr><td>Area code(s)</td><td>541</td></tr>
<tr><td>FIPS code</td><td>41-66200[1]</td></tr>
<tr><td>GNIS feature ID</td><td>1149228[2]</td></tr>
</table>

Seneca is a city in Grant County, Oregon, United States. It is located in the Blue Mountains about 23 miles south of Canyon City, on U.S. Route 395, on the edge of the Malheur National Forest.[3] The population was 223 at the 2000 census.

History

Seneca post office was established in 1895 and named by postmaster Minnie Southworth for her brother-in-law, prominent Portland judge Seneca Smith.[4] [5]

While early homesteaders moved into the valley in the late 1800s, Seneca only began growing in the 1929 when it became the northern terminus of the now-vacated Oregon and Northwestern Railroad, owned by the Edward Hines Lumber Company, which extended south to Burns.[5] At that time, large-scale shipping of Ponderosa Pine logs from Seneca and the surrounding national forest began to the Hines sawmill in Hines.[5] The company established a planing mill and railroad shops in Seneca, and it became essentially a company town.[5] In 1940 Seneca's population was 275.[6] Logging in the area began to decline in the 1970s, and the Hines company ceased operations of its lumber mills and railroad in 1984.[5] The town was incorporated as a city in 1970 as lumber company control began to wane.[5] According to the 1980 census, Seneca's population was 285.[5] The estimated population in 2007 was 270.[5]

Geography

According to the United States Census Bureau, the city has a total area of 0.8 square miles (2.1 km^2), all of it land.[7]

Situated at the confluence of Bear Creek and the Silvies River, Seneca is in Bear Valley at the northern edge of the Great Basin.[3] Seneca experiences the coolest weather in Grant County and has the distinction of the coldest official temperature recorded in the Oregon: 54 degrees below zero in 1933.[5] [8]

Demographics

As of the census[1] of 2000, there were 223 people, 95 households, and 64 families residing in the city. The population density was 267.8 people per square mile (103.7/km^2). There were 115 housing units at an average density of 138.1 per square mile (53.5/km^2). The racial makeup of the city was 99.10% White, 0.45% Asian, and 0.45% from two or more races. Hispanic or Latino of any race were 0.90% of the population.

There were 95 households out of which 29.5% had children under the age of 18 living with them, 58.9% were married couples living together, 6.3% had a female householder with no husband present, and 32.6% were non-families. 29.5% of all households were made up of individuals and 10.5% had someone living alone who was 65 years of age or older. The average household size was 2.35 and the average family size was 2.89.

In the city the population was spread out with 26.0% under the age of 18, 4.9% from 18 to 24, 25.1% from 25 to 44, 31.4% from 45 to 64, and 12.6% who were 65 years of age or older. The median age was 41 years. For every 100 females there were 110.4 males. For every 100 females age 18 and over, there were 96.4 males.

The median income for a household in the city was $27,333, and the median income for a family was $26,806. Males had a median income of $30,000 versus $23,750 for females. The per capita income for the city was $12,636. About 24.0% of families and 19.8% of the population were below the poverty line, including 22.5% of those under the age of eighteen and none of those sixty five or over.

Economy

The main industries in Seneca are cattle ranching and timber production.[6]

Education

Seneca is served by the Grant County School District (formerly the John Day School District). Seneca School is a K-8 school. After 8th grade students attend Grant Union High School in John Day. Seneca School was built in the 1930s to serve the employees of the Hines Lumber Company.[9] As of 2011, the school had 58 students.[9]

References

[1] "American FactFinder" (http://factfinder.census.gov). United States Census Bureau. . Retrieved 2008-01-31.

[2] "US Board on Geographic Names" (http://geonames.usgs.gov). United States Geological Survey. 2007-10-25. . Retrieved 2008-01-31.

[3] *Oregon Atlas & Gazetteer* (7th ed.). Yarmouth, Maine: DeLorme. 2008. p. 78. ISBN 0-89933-347-8.

[4] McArthur, Lewis A.; McArthur, Lewis L. (2003) [First published 1928]. *Oregon Geographic Names* (7th ed.). Portland, Oregon: Oregon Historical Society Press. p. 860. ISBN 9780875952772. OCLC 53075956.

[5] Engeman, Richard H. (2009). *The Oregon Companion: An Historical Gazetteer of The Useful, The Curious, and The Arcane*. Portland, Oregon: Timber Press. pp. 337–338. ISBN 978-0-88192-899-0.

[6] Writers' Program of the Work Projects Administration in the State of Oregon (1940). *Oregon: End of the Trail* (http://www.archive.org/details/oregonendoftrail00writrich). American Guide Series. Portland, Oregon: Binfords & Mort. pp. 421–422. OCLC 4874569. .

[7] "US Gazetteer files: 2010, 2000, and 1990" (http://www.census.gov/geo/www/gazetteer/gazette.html). United States Census Bureau. 2011-02-12. . Retrieved 2011-04-23.

[8] http://ggweather.com/climate/extremes_us.htm, Retrieved 2010-11-25.

[9] "Seneca School" (http://www.grantesd.k12.or.us/Seneca/index.htm). Grant County School District. . Retrieved 2011-01-17.

Further reading

- Friedman, Ralph (1978). "Sawdust at Seneca" (http://books.google.com/books?id=XK9G3wcnmrUC&printsec=frontcover&dq=Ralph+Friedman&cd=1#v=onepage&q=Seneca&f=false). *Tracking Down Oregon*. Caldwell, Idaho: The Caxton Printers, Ltd. pp. 251–2532. ISBN 0-87004-257-2.

External links

- Entry for Seneca (http://bluebook.state.or.us/local/cities/sy/seneca.htm) in the *Oregon Blue Book*
- Seneca Community Profile (http://www.orinfrastructure.org/profiles/Seneca/) from the Oregon Business Development Commission
- Seneca History Project (http://www.senecakids.org/SenecaProject/index.html) from Seneca School
- Historic images of Seneca (http://photos.salemhistory.net/cdm4/results.php?CISOOP1=all&CISOBOX1=Seneca&CISOFIELD1=CISOSEARCHALL&CISOROOT=all) from Salem Public Library

Grant_County,_Oregon

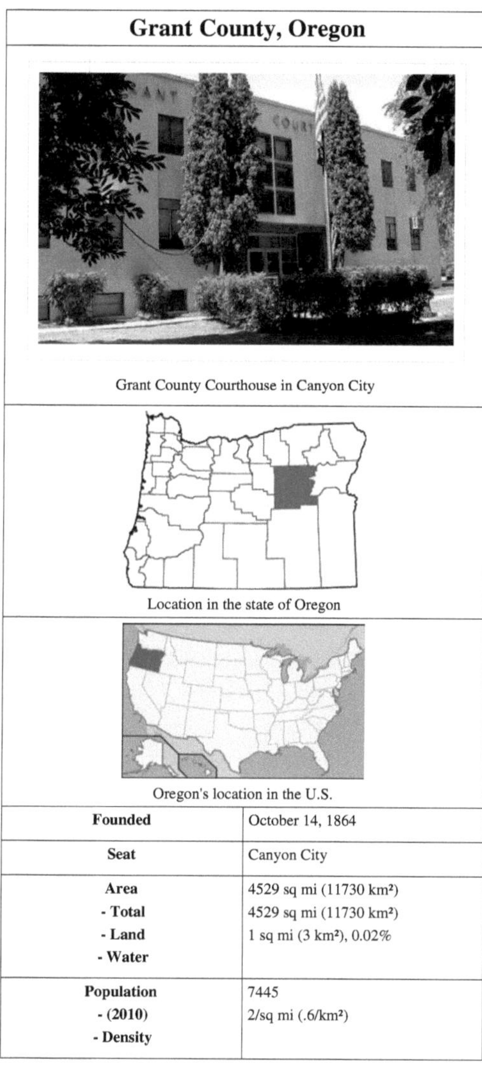

Grant County Courthouse in Canyon City

Location in the state of Oregon

Oregon's location in the U.S.

Founded	October 14, 1864
Seat	Canyon City
Area **- Total** **- Land** **- Water**	4529 sq mi (11730 km²) 4529 sq mi (11730 km²) 1 sq mi (3 km²), 0.02%
Population **- (2010)** **- Density**	7445 2/sq mi (.6/km²)

Grant County is a county located in the U.S. state of Oregon. It is included in the 8 county definition of Eastern Oregon. In 2010, its population was 7,445. It is named for President Ulysses S. Grant, who served as an army officer in the Oregon Territory, and at the time of the county's creation was a Union general in the American Civil War. The seat of the county is Canyon City.

History

Grant County was established on October 14, 1864, from parts of old Wasco and old Umatilla counties. Prior to its creation, cases brought to court were tried in The Dalles, county seat of the vast Wasco County. The great distance to The Dalles made law enforcement a difficult problem, and imposed a heavy burden on citizens who had a need to transact business at the courthouse. In 1889, more than half of the southern part of the original Grant County was taken to form Harney County. Also in 1899, a small part of northwestern Grant County was taken (along with parts of Crook and Gilliam counties) to form Wheeler County.

After gold was discovered in 1862 on Whiskey Flat, it has been estimated that within ten days 1,000 miners were camped along Canyon Creek. This increased population created a need for county government. Grant County's government operates in accordance with the Oregon Constitution which was ratified by the People of Oregon in November 1857, and the revised Statutes of Oregon. It employs the old-western county government system: the County Court, with a County Judge and two Commissioners. While the County Court no longer exercises much judicial authority, it serves as the executive branch of county government. There are no parishes or villages in Grant County, and while the term "town" is often used locally to describe one of the incorporated cities, surveyed townships have nothing to do with political divisions or organization in Oregon.

The third man to serve as County Judge of Grant County was Cincinnatus Hiner "Joaquin" Miller (1837–1913), the noted poet, playwright, and western naturalist, called the "Poet of the Sierras" and the "Byron of the Rockies."

The county seat is Canyon City, which served as the chief community of the county for many years. In 1864, when the county was organized, Canyon City is said to have boasted the largest population of any community in Oregon. Mining and ranching, along with timber and then the service and public works that followed, brought people into the area and communities grew around the natural centers of industry and agriculture. Since the 1930s, the city of John Day has served as the main economic center of the county, and boasts the largest population.

Geography

According to the U.S. Census Bureau, the county has a total area of 4529 square miles (11730 km^2), of which 4529 square miles (11730 km^2) is land and 1 square mile (2.6 km^2) (0.02%) is water.

Grant shares boundaries with more counties (eight) than any other county in Oregon.

Approximately 63% of the land area of the county is controlled by the Federal Government, most of which is controlled by the U.S. Forest Service, and the Bureau of Land Management. Grant County contains most of the Malheur National Forest and sections of the Wallowa–Whitman, Umatilla and Ochoco National Forests, and has more than 150000 acres (610 km^2) of federally-designated Wilderness Areas.

Grant County contains the headwaters of the John Day River, which has more miles of Wild and Scenic River designation than any other river in the United States.

The elevation of the county varies from 1,820 on the John Day River near Kimberly, to 9038 feet (2755 m) at the summit of Strawberry Mountain. The terrain of the county varies from grassland steppes and rangelands in relatively open or rolling hills and valleys, to steep, rugged, rocky high-alpine landscapes. Between these, the county contains heavily timbered land, many rolling hills, canyons and mountainous terrain. Portions of the county are technically high desert, dominated by sagebrush and sparse grasses.

Grant County includes the southern part of the Blue Mountains. One unique characteristic of the typical forestland of the area is the relatively low density of underbrush. Travelers and emigrants of the 19th century remarked that the absences of underbrush, and the wide spacing of the trees, made it possible to drive a wagon and team of horses virtually anywhere the grade would permit. The forested land of the county vary from sparse stands of Western Juniper in more arid, open, or rocky ground, to Sub-Alpine and High-Alpine fir stands in the highest terrain. Other forested areas (mainly above 3200 feet (980 m) in elevation) are marked by stands of Ponderosa Pine, Douglas Fir, White

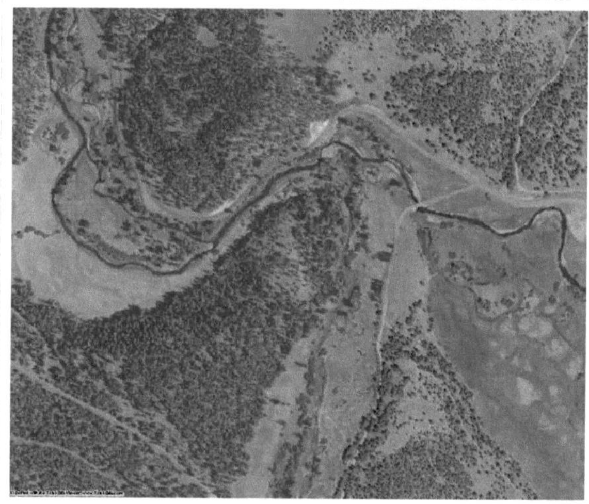

An aerial view of Grant County

Fir, Western Larch (a deciduous conifer commonly called "Tamarack"), Lodgepole Pine, Spruce stands in some higher elevation sites and a few stands of White Pine, as well as Cottonwood trees along some rivers and streams, and Birch and Quaking Aspen groves, mainly at higher elevations. There is also a rare and isolated stand of Alaskan Yellow Cedar in the Aldrich Mountains. Other flora includes a wide variety of native grasses and wildflowers, huckleberries, wild strawberries, elderberries, several types of edible mushrooms and Oregon Grape, the state plant. Non-native Russian Cheatgrass is also prevalent in many areas of the county.

Grant County is also home to what may be one of the largest living organism in the world, a giant fungus of the species *Armillaria solidipes* that lives within the Malheur National Forest. It was found to span 8.9 square kilometres (2200 acres). Its total mass has been estimated to be between 8,500 and 10,500 tons, and its age at somewhere between 2,000 and 8,500 years.[1]

The physical terrain one encounters today is far different than in prehistoric times. Fossil records show that, in the Paleozoic and early Mesozoic eras, much of the county was an ancient seabed. After emerging, the absence of the Cascade Mountains allowed the region to experience a relatively wet temperate climate. Ancient Tertiary rivers flowed through the area on courses that would be impossible today. During the Cenozoic Era, volcanic activity and extensive lava flows in the region dramatically changed the landscape. The John Day Fault (one of the only major faults in North America to run east-west) runs along the southern edge of the John Day Valley, caused an uplift, forming the Strawberry and Aldrich mountain ranges and the northern boundary of the Great Basin. Relatively recently in geological terms, during the last Ice age and shortly thereafter, large lakes were present in southeastern Oregon. Continual glaciers were still clinging to mountains in the area in the late 19th century, and one small glacier on Strawberry Mountain often remains year-round.

The geology of Grant County is rich, including one of the largest fossil concentrations in North America: The John Day Fossil Beds, which the U.S. Congress designated as a National Monument in 1974. Valuable metals, including gold, silver, platinum group elements, chrome, copper and cobalt, are found in the region. It was this mineral wealth, and the development of gold mines in particular, that spurred the permanent settlement of the area. Large zones of serpentine, a very ancient metamorphic rock (among the oldest on earth), dating from the early Mesozoic (Triassic) Era, are found in numerous locations. Strawberry Mountain (an extinct volcano), the Granite peaks and boulders of the Elkhorn Mountains, and numerous rim rocks, lava flows and Basalt outcrops are evidence of the historic volcanic activity in the region. Hydrothermal resources are still present, with a number of hot and warm springs.

The remnants of ferns, semi-tropical and temperate deciduous forests, shellfish, saber-toothed tigers, extinct horse and camel species, and giant sloth, among other extinct species found in the John Day Fossil Beds, are a reminder that the flora and fauna of the region has changed significantly over the millennia. While deer, elk, pronghorn, cougar, bear and upland game bird populations thrive today, some of these animals were remarkably scarce 200 years ago. Explorers and trappers traveling through the region in the early 19th century remarked on the scarcity of game animals and their ability (or inability, as the case were) to find food.

Native fish in the region include several trout species; warm water fish such as Bass and Perch are found in the lower John Day River; and migratory Salmon and Steelhead are found in the county seasonally. While Salmon and Steelhead returns to the John Day Basin experienced a sharp decline during the past 50 years, mainly due to the construction of large dams on the Columbia River, the major watercourses of John Day Basin remain free of physical obstructions, and the numbers of returning Salmon and Steelhead have improved in recent years, marking some of the best fish runs recorded in the past half-century.

Most of Grant County is drained by the four forks of the John Day River, all of which have their headwaters in the county. The John Day River system drains some 7900 square miles (20000 km^2). It is the third longest free-flowing river in the "lower 48" and has more miles of federal "Wild and Scenic River" designation than any other river in the United States. The river system in Grant County includes the upper 100 miles (160 km) of the Main Stem, all of the 112 miles (180 km) of the North Fork, all 75 miles (121 km) of the Middle Fork, and all 60 miles (97 km) of the South Fork of the John Day River. From Grant County, the lower John Day River flows another 184 miles (296 km) to its confluence with the Columbia River. The southeastern corner of the county includes the headwaters of the Malheur and Little Malheur rivers, which find their way to the Snake River. The southern part of Grant County includes the northern-most reaches of the Great Basin, including the Silvies River watershed, which flows south into Harney Lake in the High Desert of Eastern Oregon. A small area in the southwestern corner of Grant County is in the Crooked River and Deschutes River watersheds.

Grant County is an arid to temperate region, with average annual precipitation ranging from 9 inches (230 mm) near Picture Gorge, to over 40 inches (1000 mm) in the Strawberry Mountains. Annual precipitation in the valleys averages between 12 and 14 inches (360 mm), while the uplands or highlands of the county average between 16 and 24 inches (610 mm). Grant County averages between 40 and 60 days each year that see more than 0.10 inches (2.5 mm) of precipitation. A great deal of the county's precipitation comes in the form of winter snow in the mountains. This snow pack is vital to recharge aquifers, resulting in spring run-off, and in-stream flows of water throughout the year.

Average temperatures in the county range from the warmest community, Monument, with average daily highs/lows of 90°/50 °F in July and 42°/22 °F in January; to the coolest community, Seneca, with average daily highs/lows of 80°/38 °F in July and 33°/8 °F in January. Extreme temperatures in the county show 30-year highs/lows of: 103°/-37 °F at Austin; 112°/-23 °F at John Day; 108°/-25 °F at Long Creek; 112°/-26 °F at Monument; and 100°/-48 °F at Seneca.

Grant County has an estimated 200 days of clear sunny or mostly sunny days, or an estimated 300 days of clear sunny, mostly sunny, or partly sunny days each year. The county experiences an estimated 65 days of overcast skies, with about 165 days of partly to mostly cloudy days annually.

Economy

With the discovery of gold near Canyon City in June 1862, and near Granite in July 1862, gold miners streamed into the area. The eminent geologist, Waldemar Lindgren, who visited the area in 1900, estimated that approximately $16 million in gold had been mined from the Canyon City area alone by that time. (In 1900, the value of gold was fixed at $20.67 per ounce, so that $16 million in gold would have been roughly 800,000 ounces.) Mining remained the dominant sector of the area's economy, with increasing lode-ore production annually, until October 1942 when the U. S. War Labor Board made gold mining illegal by Executive Order, Public Law L-208. This effectively led to several mining towns being abandoned and the demise of the mining industry in eastern Oregon and elsewhere; idle equipment was removed as scrap drives during World War II literally dismantled a great deal of the county's mining infrastructure. In Oregon, Grant County's gold production was second only to Baker County.

Because of the wealth of natural resources found in Grant County, agriculture, ranching, and timber industries naturally grew with and contributed to the development of the county. In the early days, sheep formed a large part of the agricultural base and the area boasted some of the largest sheep bands in the world, supplying a great volume of wool to, among others, the world-renown Pendleton Wool Works in Pendleton. Cattle ranchers and sheep ranchers were often at odds and physical confrontations were not uncommon. By the 1920 and 1930s, however, cattle ranching became—and continues to be—the dominant sector of the agricultural industry. Crop farming, dairy production and orchards operated on small scales during the late 19th century and early 20th century, but declined during World War II due to changing market and labor pressures. The commercial timber industry in Grant County grew rapidly in the 1920s, and again during and after World War II. Livestock raising and timber harvesting remain important sectors of Grant County's economy, although the production and profitability of these industries has declined in recent years due mainly to political and expanding-market factors. Two wood-fired co-gen electric plants have been built in the county, one of which continues to operate in Prairie City.

Due mainly to federal land management policies and global market pressures affecting timber and agricultural production and extraction, the county has experienced the second highest unemployment rate in Oregon for more than 30 years. The county has experienced some growth in recreational activities (including hunting) and tourism, as well as cottage industry, but residents have struggled to develop new productive industries and to diversify their economy. Slightly more than a quarter of the county's workforce is employed by some level of government or public services.

Politics

Like all counties in eastern Oregon, the majority of registered voters who are party of a political party in Grant County are members of the Republican Party. In the 2008 presidential election, 70.97% of Grant County voters voted for Republican John McCain, while 26.05% voted for Democrat Barack Obama and 3.94% of voters either voted for a Third Party candidate or wrote in a candidate.[2] These numbers show a small but definite shift towards the Democratic candidate when compared to the 2004 presidential election, in which 78.9% of Grant County voters voted for George W. Bush, while 19.2% voted for John Kerry, and .9% of voters either voted for a Third Party candidate or wrote in a candidate.[3]

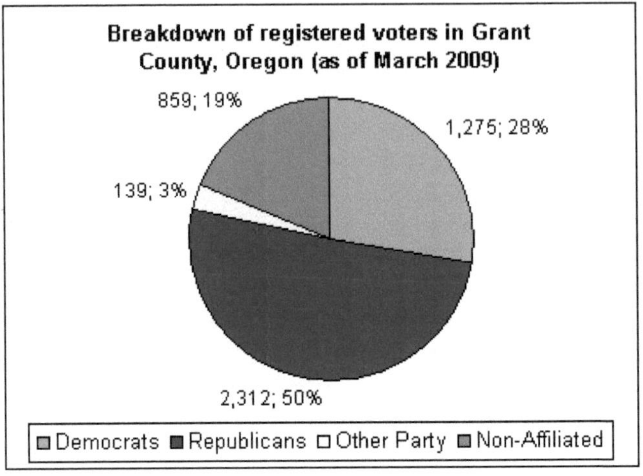

[4]

Adjacent counties

- Malheur County, Oregon - (southeast)
- Harney County, Oregon - (south)
- Crook County, Oregon - (west)
- Wheeler County, Oregon - (west)
- Morrow County, Oregon - (north)
- Umatilla County, Oregon - (north)
- Union County, Oregon - (northeast)
- Baker County, Oregon - (east)

National protected areas

- John Day Fossil Beds National Monument (part)
- Malheur National Forest (part)
- Ochoco National Forest (part)
- Umatilla National Forest (part)
- Wallowa–Whitman National Forest (part)

Demographics

Historical populations		
Census	Pop.	%±
1870	2251	—
1880	4303	91.2%
1890	5080	18.1%
1900	5948	17.1%
1910	5607	−5.7%
1920	5496	−2.0%
1930	5940	8.1%
1940	6380	7.4%
1950	8329	30.5%
1960	7726	−7.2%
1970	6996	−9.4%
1980	8210	17.4%
1990	7853	−4.3%
2000	7935	1.0%
2010	7445	−6.2%
[5] [6] [7]		

As of the census[8] of 2000, there were 7,935 people, 3,246 households, and 2,233 families residing in the county. The population density was 2 people per square mile (1/km²). There were 4,004 housing units at an average density of 1 per square mile (0/km²). The racial makeup of the county was 95.69% White, 0.10% Black or African American, 1.60% Native American, 0.19% Asian, 0.04% Pacific Islander, 0.68% from other races, and 1.70% from two or more races. 2.05% of the population were Hispanic or Latino of any race. 17.5% were of English, 17.1% German, 14.3% American and 9.0% Irish ancestry according to Census 2000.

There were 3,246 households out of which 30.10% had children under the age of 18 living with them, 57.90% were married couples living together, 7.90% had a female householder with no husband present, and 31.20% were non-families. 27.10% of all households were made up of individuals and 10.90% had someone living alone who was 65 years of age or older. The average household size was 2.39 and the average family size was 2.89.

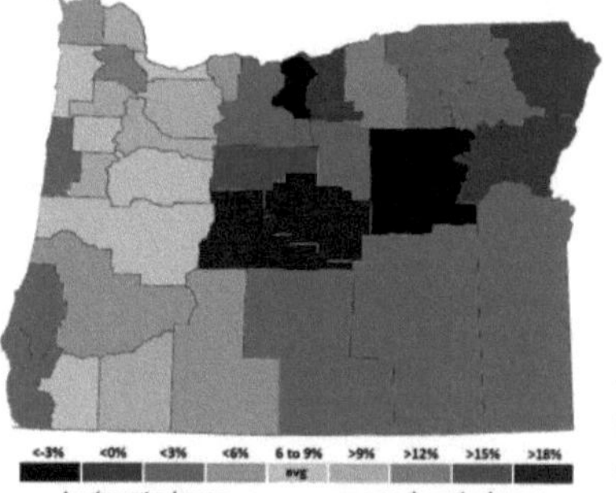

From 2000 to 2007, Grant County lost 4.5% of its population, more than any county in the state.

In the county, the population was spread out with 25.80% under the age of 18, 5.60% from 18 to 24, 24.00% from 25 to 44, 27.90% from 45 to 64, and 16.80% who were 65 years of age or older. The median age was 42 years. For every 100 females there were 99.30 males. For every 100 females age 18 and over, there were 97.10 males.

The median income for a household in the county was $32,560, and the median income for a family was $37,159. Males had a median income of $31,843 versus $22,253 for females. The per capita income for the county was

$16,794. About 11.20% of families and 13.70% of the population were below the poverty line, including 16.60% of those under age 18 and 10.20% of those age 65 or over.

2000 U.S. Census statistics for Grant County show that the total workforce for Grant County was 3,800, or 62% of the total population over age 16. These people were employed as follows:

56.9% private wage/salaried positions; 14.7% private self-employed (not incorporated business); 0.8% private unpaid family workers; 27.6% public employees (municipal, county, state, federal governments);

By industry: 20.6% education, health, social services; 17.3% agriculture, forestry, mining; 10.0% manufacturing; 9.8% retail trade; 7.6% arts, entertainment, recreation, accommodations, and food; 6.9% public administration; 6.5% construction; 5.9% other services; 5.1% transportation, warehousing, utilities; 4.1% professional, administrative, and waste management; 3.1% finance, insurance, real estate, leasing; 1.7% information; 1.5% wholesale trade;

Communities

Incorporated cities

- Canyon City
- Dayville
- Granite
- John Day
- Long Creek
- Monument
- Mount Vernon
- Prairie City
- Seneca

Unincorporated communities and CDPs

- Austin
- Austin Junction
- Bates
- Beech Creek
- Cabell City
- Courtrock
- Dale
- Fox
- Galena
- Hamilton
- Izee
- Kimberly
- Range
- Ritter
- Robinsonville
- Silvies
- Susanville
- Three Forks

See also

- National Register of Historic Places listings in Grant County, Oregon

References

[1] http://www.abc.net.au/cgi-bin/common/printfriendly.pl?/science/news/enviro/EnviroRepublish_828525.htm *Humungous fungus: world's largest organism?* Retrieved on 5/09/011

[2] http://www.gcoregonlive2.com/votes2008.php Retrieved on 4/21/09

[3] http://www.city-data.com/county/Grant_County-OR.html Retrieved on 4/21/09

[4] http://www.sos.state.or.us/elections/votreg/mar09.pdf Retrieved on 4/21/09

[5] http://www.census.gov/population/www/censusdata/cencounts/files/or190090.txt

[6] http://factfinder2.census.gov

[7] http://mapserver.lib.virginia.edu/

[8] "American FactFinder" (http://factfinder.census.gov). United States Census Bureau. . Retrieved 2008-01-31.

External links

* Grant County Chamber of Commerce (http://www.gcoregonlive.com/)

Oregon

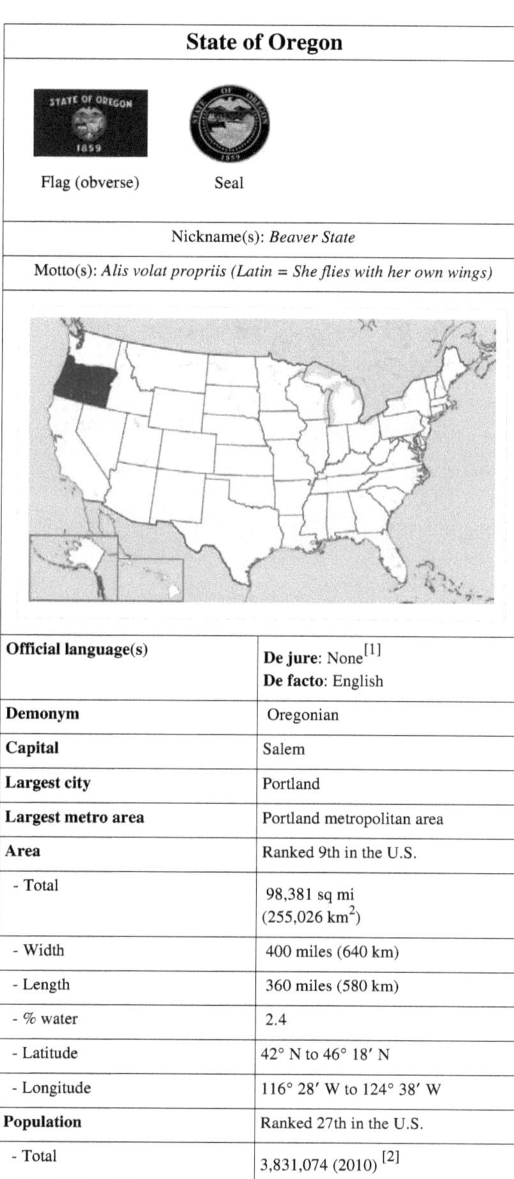

State of Oregon	
Flag (obverse)	Seal
Nickname(s): *Beaver State*	
Motto(s): *Alis volat propriis (Latin = She flies with her own wings)*	
Official language(s)	**De jure**: None[1] **De facto**: English
Demonym	Oregonian
Capital	Salem
Largest city	Portland
Largest metro area	Portland metropolitan area
Area	Ranked 9th in the U.S.
- Total	98,381 sq mi (255,026 km^2)
- Width	400 miles (640 km)
- Length	360 miles (580 km)
- % water	2.4
- Latitude	42° N to 46° 18′ N
- Longitude	116° 28′ W to 124° 38′ W
Population	Ranked 27th in the U.S.
- Total	3,831,074 (2010) [2]

- Density	39.9[3] /sq mi (15.41/km^2) Ranked 39th in the U.S.
Elevation	
- Highest point	Mount Hood[4] [5] [6] 11,249 ft (3,428.8 m)
- Mean	3,300 ft (1,000 m)
- Lowest point	Pacific Ocean[5] sea level
Admission to Union	February 14, 1859 (33rd)
Governor	John Kitzhaber (D)
Secretary of State	Kate Brown (D)
Legislature	Legislative Assembly
- Upper house	State Senate
- Lower house	House of Representatives
U.S. Senators	Ron Wyden (D) Jeff Merkley (D)
U.S. House delegation	4 Democrats, 1 Republican (list)
Time zones	
- most of state	Pacific: UTC-8/-7
- most of Malheur County	Mountain: UTC-7/-6
Abbreviations	OR Ore. US-OR
Website	[www.oregon.gov www.oregon.gov]

Oregon (🔊 [i]/ˈɒrɪgən/ *orr-ə-gən*)[8] is a state in the Pacific Northwest region of the United States. It is located on the Pacific coast, with Washington to the north, California to the south, Nevada on the southeast and Idaho to the east. The Columbia and Snake rivers delineate much of Oregon's northern and eastern boundaries, respectively. The area was inhabited by many indigenous tribes before the arrival of traders, explorers, and settlers who formed an autonomous government in Oregon Country in 1843. The Oregon Territory was created in 1848, and Oregon became the 33rd state on February 14, 1859.

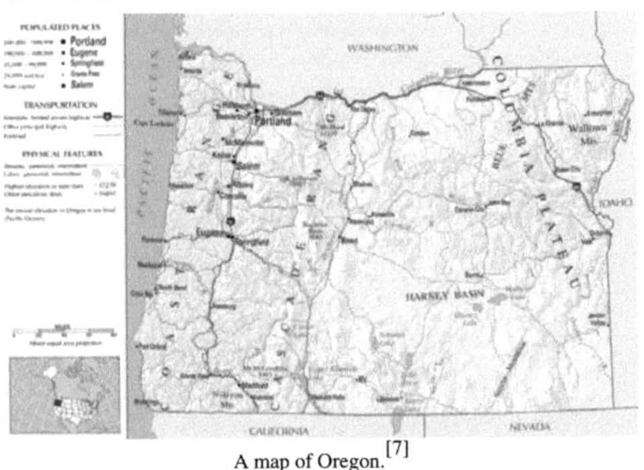

A map of Oregon.[7]

Salem is the state's capital and third-most-populous city; Portland is the most populous. Oregon's 2010 population is just over 3.8 million, a 12% increase over 2000.[9] Portland is the 29th-largest U.S. city, with a population of 583,776 (2010 US Census) and a metro population of 2,241,841 (2009 estimate), the 23rd-largest U.S. metro area. The valley of the Willamette River in western Oregon is the state's most densely populated area and is home to eight of the ten most populous cities.

The Oregon coastline looking south from Ecola State Park, with Haystack Rock in the distance.

Oregon contains a diverse landscape including the windswept Pacific coastline, the volcanoes of the rugged and glaciated Cascade Mountain Range, many waterfalls (including Multnomah Falls), dense evergreen forests, mixed forests and deciduous forests at lower elevations, and high desert across much of the eastern portion of the state, extending into the Great Basin. The tall Douglas firs and redwoods along the rainy Western Oregon coast contrast with the lower density and fire prone pine tree and juniper forests covering portions of the eastern half of the state. Alder trees are common in the west and fix nitrogen for the conifers and aspen groves are common in eastern Oregon. Stretching east from Central Oregon, the state also includes semi-arid shrublands, prairies, deserts, steppes, and meadows. Mount Hood is the highest point in the state at 11249 feet (3429 m). Crater Lake National Park is the only national park in Oregon.

History

Humans have inhabited the area that is now Oregon for at least 15,000 years. In recorded history, mentions of the land date to as early as the 16th century. During the 18th and 19th centuries, European powers—and later the United States—quarreled over possession of the region until 1846 when the U.S. and Great Britain finalized division of the region. Oregon became a state in 1859 and is now home to over 3.8 million residents.

Earliest inhabitants

Human habitation of the Pacific Northwest began at least 15,000 years ago, with the oldest evidence of habitation in Oregon found at Fort Rock Cave and the Paisley Caves in Lake County. Archaeologist Luther Cressman dated material from Fort Rock to 13,200 years ago.[10] By 8000 B.C. there were settlements throughout the state, with populations concentrated along the lower Columbia River, in the western valleys, and around coastal estuaries.

European exploration

By the 16th century Oregon was home to many Native American groups, including the Coquille (Ko-Kwell),Bannock, Chasta, Chinook, Kalapuya, Klamath, Molalla, Nez Perce, Takelma, and Umpqua.[11] [12] [13] [14]

The first Europeans to visit Oregon were Spanish explorers who sighted southern Oregon off the Pacific Coast in 1543. No Europeans returned to Oregon until 1778, when British captain James Cook explored the coast.[15] French Canadian and metis trappers and missionaries arrived in the eastern part of the state in the late 18th century and early 19th century, many having travelled as members of Lewis and Clark and the 1811 Astor expeditions. Some stayed permanently, including Étienne Lussier, believed to be the first European farmer in the state of Oregon. The evidence of this French Canadian presence can be found in the numerous names of French origin in that part of the state: Charbonneau, Malheur Lake and River, Grande Ronde and Des Chutes Rivers, city of Ontario, etc.

During U.S. westward expansion

The Lewis and Clark Expedition traveled through the region also in search of the Northwest Passage. They built their winter fort at Fort Clatsop, near the mouth of the Columbia River. British explorer David Thompson also conducted overland exploration.

In 1811, David Thompson, of the North West Company, became the first European to navigate the entire Columbia River. Stopping on the way, at the junction of the Snake River, he posted a claim to the region for Great Britain and the North West Company. Upon returning to Montreal, he publicized the abundance of fur-bearing animals in the area.

Also in 1811, New Yorker John Jacob Astor financed the establishment of Fort Astoria at the mouth of the Columbia River as a western outpost to his Pacific Fur Company;[16] this was the first permanent European settlement in Oregon.

In the War of 1812, the British gained control of all Pacific Fur Company posts. The Treaty of 1818 established joint British and American occupancy of the region west of the Rocky Mountains to the Pacific Ocean. By the 1820s and 1830s, the Hudson's Bay Company dominated the Pacific Northwest from its Columbia District headquarters at Fort Vancouver (built in 1825 by the District's Chief Factor John McLoughlin across the Columbia from present-day Portland).

In 1841, the expert trapper and entrepreneur Ewing Young died leaving considerable wealth and no apparent heir, and no system to probate his estate. A meeting followed Young's funeral at which a probate government was proposed. Doctor Ira Babcock of Jason Lee's Methodist Mission was elected Supreme Judge. Babcock chaired two meetings in 1842 at Champoeg (half way between Lee's mission and Oregon City) to discuss wolves and other animals of contemporary concern. These meetings were precursors to an all-citizen meeting in 1843, which instituted a provisional government headed by an executive committee made up of David Hill, Alanson Beers, and Joseph Gale. This government was the first acting public government of the Oregon Country before annexation by the government of the United States.

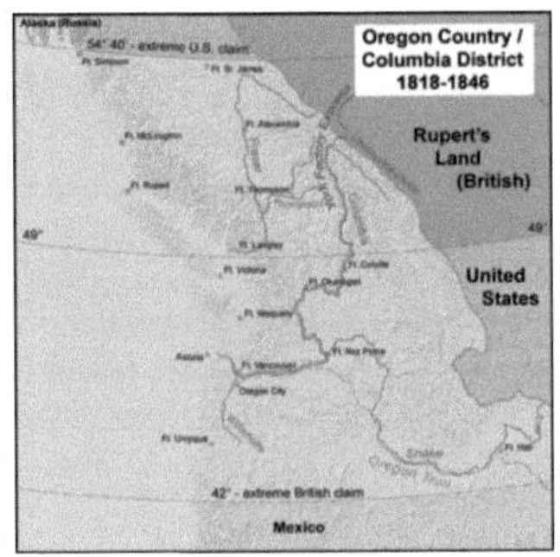

Map of Oregon Country.

Also in 1841, Sir George Simpson, Governor of the Hudson's Bay Company, reversed the Hudson's Bay Company's long-standing policy of discouraging settlement because it interfered with the lucrative fur trade. He directed that some 200 Red River Colony settlers be relocated to HBC farms near Fort Vancouver, (the James Sinclair expedition), in an attempt to hold Columbia District.

Starting in 1842–1843, the Oregon Trail brought many new American settlers to Oregon Country. For some time, it seemed that Britain and the United States would go to war for a third time in 75 years (see Oregon boundary dispute), but the border was defined peacefully in 1846 by the Oregon Treaty. The border between the United States and British North America was set at the 49th parallel. The Oregon Territory was officially organized in 1848.

Settlement increased with the Donation Land Claim Act of 1850 and the forced relocation of the native population to Indian reservations in Oregon.

After statehood

Oregon was admitted to the Union on February 14, 1859. Founded as a refuge from disputes over slavery, Oregon had a "whites only" clause in its original state Constitution.[17]

At the outbreak of the American Civil War, regular U.S. troops were withdrawn and sent east. Volunteer cavalry recruited in California were sent north to Oregon to keep peace and protect the populace. The First Oregon Cavalry served until June 1865.

In the 1880s, the growth of railroads helped market the state's lumber, wheat, and the rapid growth of its cities.

20th and 21st centuries

In 1902, Oregon introduced direct legislation by the state's citizens through initiatives and referenda, known as the Oregon System. Oregon state ballots often include politically conservative proposals side-by-side with politically liberal ones, illustrating the diversity of political thought in the state.

Industrial expansion began in earnest following the construction of the Bonneville Dam in 1933–1937 on the Columbia River. Hydroelectric power, food, and lumber provided by Oregon helped fuel the development of the West, although the periodic fluctuations in the U.S. building industry have hurt the state's economy on multiple occasions.

Name

Oregon welcome sign at Hells Canyon.

The earliest known use of the name, spelled "Ouragon", was in a 1765 petition by Major Robert Rogers to the Kingdom of Great Britain. The term referred to the then–mythical River of the West (the Columbia River). By 1778 the spelling had shifted to *Oregon*.[18] In his 1765 petition, Rogers wrote:[19]

> "The rout [*sic*]...is from the Great Lakes towards the Head of the Mississippi, and from thence to the River called by the Indians Ouragon..."

One theory is the name comes from the French word *ouragan* ("windstorm" or "hurricane"), which was applied to the River of the West based on Native American tales of powerful Chinook winds of the lower Columbia River, or perhaps from firsthand French experience with the chinook winds of the Great Plains. At the time the River of the West was thought to rise in western Minnesota and flow west through the Great Plains.[20]

Joaquin Miller explained in Sunset (magazine) in 1904 how Oregon's name was derived:[21]

> "The name, Oregon, is rounded down phonetically, from *Aure il agua*—Oragua, Or-a-gon, Oregon—given probably by the same Portuguese navigator that named the Farallones after his first officer, and it literally, in a large way, means cascades: 'Hear the waters.' You should steam up the Columbia and hear and feel the waters falling out of the clouds of Mount Hood to understand entirely the full meaning of the name *Aure il agua*, Oregon."

Another account, endorsed as the "most plausible explanation" in the book *Oregon Geographic Names*, was advanced by George R. Stewart in a 1944 article in *American Speech*. According to Stewart, the name came from an engraver's error in a French map published in the early 18th century, on which the Ouisiconsink (Wisconsin) River was spelled "Ouaricon-sint", broken on two lines with the -sint below, so there appeared to be a river flowing to the west named "Ouaricon".

According to the Oregon Tourism Commission (also known as Travel Oregon), present-day Oregonians /ˌɒrɪˈɡoʊniənz/[22] pronounce the state's name as "OR-UH-GUN, never OR-EE-GONE".[8]

After being drafted by the Detroit Lions in 2002, former Oregon Ducks quarterback Joey Harrington distributed "ORYGUN" stickers to members of the media as a reminder of how to pronounce the name of his home state.[23] [24] The stickers are sold by the University of Oregon Bookstore, which credits the spelling as a joke that is meant "for Oregonians and Oregon fans everywhere who get a kick out of this hilarious mispronunciation of our state."[25]

Geography

National parks and historic areas in Oregon

Entity	Location
Crater Lake National Park	Southern Oregon
John Day Fossil Beds National Monument	Eastern Oregon
Newberry National Volcanic Monument	Central Oregon
Cascade–Siskiyou National Monument	Southern Oregon
Oregon Caves National Monument	Southern Oregon
California Trail	Southern Oregon, California
Fort Vancouver National Historic Site	Western Oregon, Washington
Lewis and Clark National Historic Trail	IL, MO, KS, IA, NE, SD, ND, MT, ID, **OR**, WA
Lewis and Clark National and State Historical Parks	Western Oregon, Washington
Nez Perce National Historical Park	MT, ID, **OR**, WA
Oregon Trail	MO, KS, NE, WY, ID, **OR**

Oregon's geography may be split roughly into eight areas:

- Oregon Coast—west of the Coast Range
- Willamette Valley
- Rogue Valley
- Cascade Mountains
- Klamath Mountains
- Columbia River Plateau
- Oregon Outback
- Blue Mountains (ecoregion)

The mountainous regions of western Oregon, home to four of the most prominent mountain peaks of the United States including Mount Hood, were formed by the volcanic activity of Juan de Fuca Plate, a tectonic plate that poses a continued threat of volcanic activity and earthquakes in the region. The most recent major activity was the 1700 Cascadia earthquake. (Washington's Mount St. Helens erupted in 1980, an event which was visible from Oregon.)

The Columbia River, which forms much of the northern border of Oregon, also played a major role in the region's geological evolution, as well as its economic and cultural development. The Columbia is one of North America's largest rivers, and one of two rivers to cut through the Cascades (the Klamath River in Southern Oregon is the other). About 15,000 years ago, the Columbia repeatedly flooded much of Oregon during the Missoula Floods; the modern fertility of the Willamette Valley is largely a result of those floods. Plentiful salmon made parts of the river, such as Celilo Falls, hubs of economic activity for thousands of years. In the 20th century, numerous hydroelectric dams were constructed along the Columbia, with major impacts on salmon, transportation and commerce, electric power, and flood control.

Today, Oregon's landscape varies from rain forest in the Coast Range to barren desert in the southeast, which still meets the technical definition of a frontier.

Oregon is 295 miles (475 km) north to south at longest distance, and 395 miles (636 km) east to west at longest distance. In land and water area, Oregon is the ninth largest state, covering 98381 square miles (km^2).[26] The highest point in Oregon is the summit of Mount Hood, at 11239 feet (3426 m), and its lowest point is sea level of the Pacific Ocean along the Oregon coast.[27] Its mean elevation is 3300 feet (1006 m). Crater Lake National Park is the state's only national park and the site of Crater Lake, the deepest lake in the U.S. at 1943 feet (592 m).[28] Oregon claims the D River is the shortest river in the world,[29] though the American state of Montana makes the same claim of its Roe River.[30] Oregon is also home to Mill Ends Park (in Portland),[31] the smallest park in the world at 452 square inches (0.29 m^2). Oregon's geographical center is farther west than any of the other 48 contiguous states (although the westernmost point of the lower 48 states is in Washington).

Oregon is home to what is considered the largest single organism in the world, an *Armillaria solidipes* fungus beneath the Malheur National Forest of eastern Oregon.[32]

Images of Oregon

Mount Hood, with Trillium Lake in the foreground

An aerial view of Crater Lake in Oregon

Sunset over Malheur Butte, an extinct volcanic cinder cone near Ontario, Oregon.

Portland

Map of Oregon's population density

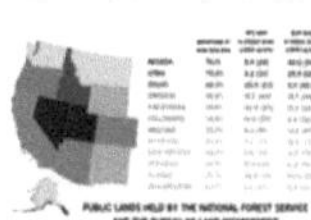

Nearly half of Oregon's land is held by the U.S. Forest Service and the Bureau of Land Management.[33]

Willamette Valley basin

Major cities

Most Populous Cities

City	Population (2010 US Census)[34]
1. Portland	583,776
2. Eugene	156,185
3. Salem	154,637
4. Gresham	105,594
5. Hillsboro	91,611
6. Beaverton	89,803
7. Bend	76,639
8. Medford	74,907
9. Springfield	59,403
10. Corvallis	54,462
11. Albany	50,158
12. Tigard	48,035

Further information: List of cities and unincorporated communities in Oregon

Oregon's population is largely concentrated in the Willamette Valley, which stretches from Eugene in the south (home of the University of Oregon) through Corvallis (home of Oregon State University) and Salem (the capital) to Portland (Oregon's largest city).[34]

Astoria, at the mouth of the Columbia River, was the first permanent English-speaking settlement west of the Rockies in what is now the United States. Oregon City, at the end of the Oregon Trail, was the Oregon Territory's first incorporated city, and was its first capital from 1848 until 1852, when the capital was moved to Salem. Bend, near the geographic center of the state, is one of the ten fastest-growing metropolitan areas in the United States.[35] In the southern part of the state, Medford is a rapidly growing metro area, which is home to The Rogue Valley International-Medford Airport, the third-busiest airport in the state. To the south, near the California-Oregon border, is the community of Ashland, home of the Tony Award-winning Oregon Shakespeare Festival.

Climate

Oregon's climate—particularly in the western part of the state—is heavily influenced by the Pacific Ocean. The climate is mild, but periods of extreme hot and cold can affect parts of the state. Oregon's population centers, which lie mostly in the western part of the state, are moist and mild, while the lightly populated high deserts of Central and Eastern Oregon are much drier. Oregon's highest recorded temperature is 119 °F (48 °C) at Pendleton on August 10, 1898 and the lowest recorded temperature is −54 °F (−48 °C) at Seneca on February 10, 1933.

Law and government

A writer in the Oregon Country book *A Pacific Republic*, written in 1839, predicted the territory was to become an independent republic. Four years later, in 1843, settlers of the Willamette Valley voted in majority for a republic government.[36] The Oregon Country functioned in this way until August 13, 1848, when Oregon was annexed by the United States and a territorial government was established. Oregon maintained a territorial government until February 14, 1859, when it was granted statehood.[37]

The flags of the United States and Flag of Oregon flown side-by-side in downtown Portland.

State

Oregon state government has a separation of powers similar to the federal government. It has three branches, called departments by the state's constitution:

- a legislative department (the bicameral Oregon Legislative Assembly),
- an executive department which includes an "administrative department" and Oregon's governor serving as chief executive, and
- a judicial department, headed by the Chief Justice of the Oregon Supreme Court.

Governors in Oregon serve four-year terms and are limited to two consecutive terms, but an unlimited number of total terms. Oregon has no lieutenant governor; in the event that the office of governor is vacated, Article V, Section 8a of the Oregon Constitution specifies that the Secretary of State is first in line for succession.[38] The other statewide officers are Treasurer, Attorney General, Superintendent, and Labor Commissioner. The biennial Oregon Legislative Assembly consists of a thirty-member Senate and a sixty-member House. The state supreme court has seven elected justices, currently including the only two openly gay state supreme court justices in the nation. They choose one of their own to serve a six-year term as Chief Justice. The only court that may reverse or modify a decision of the Oregon Supreme Court is the Supreme Court of the United States.

The debate over whether to move to annual sessions is a long-standing battle in Oregon politics, but the voters have resisted the move from citizen legislators to professional lawmakers. Because Oregon's state budget is written in two-year increments and, having no sales tax, its revenue is based largely on income taxes, it is often significantly over- or under-budget. Recent legislatures have had to be called into special session repeatedly to address revenue shortfalls resulting from economic downturns, bringing to a head the need for more frequent legislative sessions. Oregon Initiative 71, passed in 2010, mandates the Legislature to begin meeting every year, for 160 days in odd numbered years, and 35 days in even numbered years.

The state maintains formal relationships with the nine federally recognized tribes in Oregon:

- Burns Paiute Tribe
- Confederated Tribes of Coos, Lower Umpqua and Siuslaw Indians
- Confederated Tribes of Grand Ronde
- Confederated Tribes of Siletz Indians
- Confederated Tribes of Warm Springs
- Confederated Tribes of the Umatilla Indian Reservation
- Cow Creek Band of Umpqua Tribe of Indians
- Klamath Tribes
- Coquille Indian Tribe

Oregon State Capitol

Oregonians have voted for the Democratic Presidential candidate in every election since 1988. In 2004 and 2006, Democrats won control of the state Senate and then the House. Since the late 1990s, Oregon has been represented by four Democrats and one Republican in the U.S. House of Representatives. Since 2009, the state has had two Democratic Senators, Ron Wyden and Jeff Merkley. Oregon voters have elected Democratic governors in every election since 1986, most recently electing John Kitzhaber over Republican Chris Dudley in 2010.

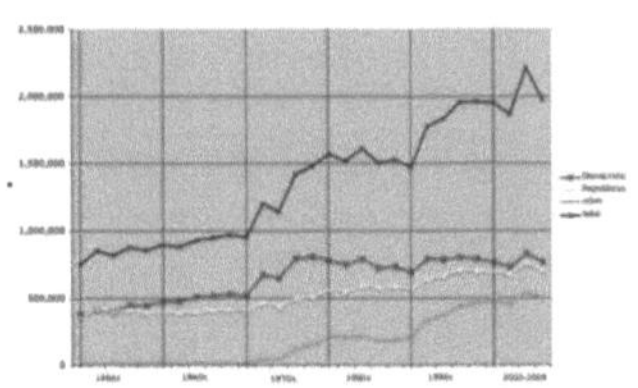

Party registration in Oregon, 1950–2006.
total Democratic Party Republican
Party non-affiliated and minor parties

The base of Democratic support is largely concentrated in the urban centers of the Willamette Valley. The eastern two-thirds of the state beyond the Cascade Mountains typically votes Republican; in 2000 and 2004, George W. Bush carried every county east of the Cascades. However, the region's sparse population means that the more populous counties in the Willamette Valley usually outweigh the eastern counties in statewide elections.

Oregon's politics are largely similar to those of neighboring Washington, for instance in the contrast between urban and rural issues.

In the 2002 general election, Oregon voters approved a ballot measure to increase the state minimum wage automatically each year according to inflationary changes, which are measured by the consumer price index (CPI).[39] In the 2004 general election, Oregon voters passed ballot measures banning same-sex marriage,[40] and restricting land use regulation.[41] In the 2006 general election, voters restricted the use of eminent domain and extended the state's discount prescription drug coverage.[42]

The distribution, sales and consumption of alcoholic beverages are regulated in the state by the Oregon Liquor Control Commission. Thus, Oregon is an Alcoholic beverage control state. While wine and beer are available in most grocery stores, few stores sell hard liquor.

In March 2011, Oregon ranked amongst the top seven "Best" states in the American State Litter Scorecard, for overall effectiveness and quality of its public space cleanliness—-primarily roadway and adjacent litter—from state and related debris removal efforts. [43]

Federal

Like all U.S. states, Oregon is represented by two U.S. Senators. Since the 1980 census, Oregon has had five Congressional districts.

After Oregon was admitted to the Union, it began with a single member in the House of Representatives (La Fayette Grover, who served in the 35th United States Congress for less than a month). Congressional apportionment increased the size of the delegation following the censuses of 1890, 1910, 1940, and 1980. A detailed list of the past and present Congressional delegations from Oregon is available.

The United States District Court for the District of Oregon hears federal cases in the state. The court has courthouses in Portland, Eugene, Medford, and Pendleton. Also in Portland is the federal bankruptcy court, with a second branch in Eugene.[44] Oregon (among other western states and territories) is in the 9th Court of Appeals. One of the court's meeting places is at the Pioneer Courthouse in downtown Portland, a National Historic Landmark built in 1869.

Politics

Presidential elections results[45]

Year	Republican	Democratic
2008	40.40% 738,475	**56.75%** *1,037,291*
2004	47.19% 866,831	**51.35%** *943,163*
2000	46.46% 713,577	**47.01%** *720,342*
1996	39.06% 538,152	**47.15%** *649,641*
1992	32.53% 475,757	**42.48%** *621,314*
1988	46.61% 560,126	**51.28%** *616,206*
1984	**55.91%** *685,700*	43.74% 536,479
1980	**48.33%** *571,044*	38.67% 456,890
1976	**47.78%** *492,120*	47.62% 490,407
1972	**52.45%** *486,686*	42.33% 392,760
1968	**49.83%** *408,433*	43.78% 358,866
1964	35.96% 282,779	**63.72%** *501,017*
1960	**52.56%** *408,060*	47.32% 367,402
1956	**55.25%** *406,393*	44.75% 329,204
1952	**60.54%** *420,815*	38.93% 270,579

The state has been thought of as politically split by the Cascade Range, with western Oregon being liberal and Eastern Oregon being conservative. In a 2008 analysis of the 2004 presidential election, a political analyst found that according to the application of a Likert scale, Oregon boasted both the most liberal voters and the most conservative voters, making it the most politically polarized state in the country.[46]

During Oregon's history it has adopted many electoral reforms proposed during the Progressive Era, through the efforts of William S. U'Ren and his Direct Legislation League. Under his leadership, the state overwhelmingly approved a ballot measure in 1902 that created the initiative and referendum for citizens to introduce or approve proposed laws or amendments to the state constitution directly, making Oregon the first state to adopt such a system. Today, roughly half of U.S. states do so.[47]

In following years, the primary election to select party candidates was adopted in 1904, and in 1908 the Oregon Constitution was amended to include recall of public officials. More recent amendments include the nation's first doctor-assisted suicide law,[48] called the Death with Dignity law (which was challenged, unsuccessfully, in 2005 by the Bush administration in a case heard by the U.S. Supreme Court), legalization of medical cannabis, and among the nation's strongest anti-urban sprawl and pro-environment laws. More recently, 2004's Measure 37 reflects a backlash against such land use laws. However, a further ballot measure in 2007, Measure 49, curtailed many of the provisions of 37.

Of the measures placed on the ballot since 1902, the people have passed 99 of the 288 initiatives and 25 of the 61 referendums on the ballot, though not all of them survived challenges in courts (see *Pierce v. Society of Sisters*, for an example). During the same period, the legislature has referred 363 measures to the people, of which 206 have passed.

Oregon pioneered the American use of postal voting, beginning with experimentation approved by the Oregon Legislative Assembly in 1981 and culminating with a 1998 ballot measure mandating that all counties conduct elections by mail. It remains the only state where voting by mail is the only method of voting.[49]

Under the leadership of Governor John Kitzhaber in 1994, Oregon was the first state in the US to set up effective health care programs with the Oregon Health Plan, which made health care available to most of its citizens without private health insurance.

In the U.S. Electoral College, Oregon casts seven votes. Oregon has supported Democratic candidates in the last six elections. Democrat Barack Obama won the state in 2008 by a margin of sixteen percentage points, with over 56% of the popular vote.

Economy

The Gross Domestic Product (GDP) of Oregon in 2010 was $168.6 billion, it is the United States's 26th wealthiest state by GDP. The state's per capita personal income in 2010 was $44,447.[50] Land in the Willamette Valley owes its fertility to the Missoula Floods, which deposited lake sediment from Glacial Lake Missoula in western Montana onto the valley floor.[51]

Oregon is also one of four major world hazelnut growing regions, and produces 95% of the domestic hazelnuts in the United States. While the history of the wine production in Oregon can be traced to before Prohibition, it became a significant industry beginning in the 1970s. In 2005, Oregon ranked third among U.S. states with 303 wineries.[52] Due to regional similarities in climate and soil, the grapes planted in Oregon are often the same varieties found in the French regions of Alsace and Burgundy. In the Southern Oregon coast commercially cultivated cranberries account for about 7 percent of U.S. production, and the cranberry ranks twenty-third among Oregon's top fifty agricultural commodities. From 2006 to 2008, Oregon growers harvested between forty and forty-nine million pounds of berries every year. Cranberry cultivation in Oregon uses about 27,000 acres in southern Coos and northern Curry counties, centered around the coastal city of Bandon, Oregon.

A grain elevator in Halsey storing grass seed, one of the state's largest crops.

In the northeastern region of the state, particularly around Pendleton, both irrigated and dry land wheat is grown. Oregon farmers and ranchers also produce cattle, sheep, dairy products, eggs and poultry.

Vast forests have historically made Oregon one of the nation's major timber production and logging states, but forest fires (such as the Tillamook Burn), over-harvesting, and lawsuits over the proper management of the extensive federal forest holdings have reduced the timber produced. According to the Oregon Forest Resources Institute, between 1989 and 2001 the amount of timber harvested from federal lands dropped some 96%, from 4,333 million to 173 million board feet (10,000,000 to 408,000 m^3), although harvest levels on private land have remained relatively constant.[53]

Even the shift in recent years towards finished goods such as paper and building materials has not slowed the decline of the timber industry in the state. The effects of this decline have included Weyerhaeuser's acquisition of Portland-based Willamette Industries in January 2002, the relocation of Louisiana-Pacific's corporate headquarters from Portland to Nashville, and the decline of former lumber company towns such as Gilchrist. Despite these changes, Oregon still leads the United States in softwood lumber production; in 2001, 6,056 million board feet (14,000,000 m^3) was produced in Oregon, compared with 4,257 million board feet (10,050,000 m^3) in Washington, 2,731 million board feet (6,444,000 m^3) in California, 2,413 million board feet (5,694,000 m^3) in Georgia, and 2,327 million board feet (5,491,000 m^3) in Mississippi.[54] The slow of the timber and lumber industry has caused high unemployment rates in rural areas.[55]

Oregon occasionally hosts film shoots. Movies filmed in Oregon include: *Rooster Cogburn, The Goonies, National Lampoon's Animal House, Stand By Me, Kindergarten Cop, Overboard, The River Wild, One Flew Over the Cuckoo's Nest, Paint Your Wagon, The Hunted, Sometimes a Great Notion, Elephant, Bandits, The Ring, The Ring Two, Quarterback Princess, The General, Mr. Brooks, Teenage Mutant Ninja Turtles III, Short Circuit, Come See the Paradise, The Shining, Drugstore Cowboy, My Own Private Idaho, The Postman, Homeward Bound, Free Willy, Free Willy 2: The Adventure Home, 1941, Swordfish, Twilight, Untraceable, Mean Creek,* and *Wendy and Lucy.* Oregon native Matt Groening, creator of *The Simpsons,* has incorporated many references from his hometown of Portland into the TV series.[56]

In late 2008, Hells Canyon and Oregon's badlands were a set location for an episode of Man vs. Wild.[57]

Largest Public Corporations Headquartered in Oregon[58]

Corporation	Headquarters	Market cap (million)
1. Nike, Inc.	near Beaverton	$32,039
2. Precision Castparts Corp.	Portland	$16,158
3. FLIR Systems	Wilsonville	$4,250
4. StanCorp Financial Group	Portland	$2,495
5. Schnitzer Steel Industries	Portland	$1,974
6. Portland General Electric	Portland	$1,737
7. Columbia Sportswear	near Beaverton	$1,593
8. Northwest Natural Gas	Portland	$1,287
9. Mentor Graphics	Wilsonville	$976
10. TriQuint Semiconductor	Hillsboro	$938

High technology industries and services have been a major employer since the 1970s. Tektronix was the largest private employer in Oregon until the late 1980s. Intel's creation and expansion of several facilities in eastern Washington County continued the growth that Tektronix had started. Intel, the state's largest for-profit private employer, operates four large facilities, with Ronler Acres, Jones Farm and Hawthorn Farm all located in Hillsboro.[59]

The spinoffs and startups that were produced by these two companies led to the establishment in that area of the so-called Silicon Forest. The recession and dot-com bust of 2001 hit the region hard; many high technology employers reduced the number of their employees or went out of business. Open Source Development Labs made news in 2004 when they hired Linus Torvalds, developer of the Linux kernel. Recently, biotechnology giant Genentech purchased several acres of land in Hillsboro to expand its production capabilities.[60] Oregon is home to several large datacenters that take advantage of cheap power and a climate in Central Oregon conducive to reducing cooling costs. Google has a large datacenter in The Dalles; Facebook is building a datacenter in Prineville; and Amazon is restarting construction of a datacenter in Boardman.

Oregon is also the home of large corporations in other industries. The world headquarters of Nike, Inc. are located near Beaverton. Medford is home to Harry and David, which sells gift items under several brands. Medford is also home to the national headquarters of the Fortune 1000 company, Lithia Motors. Portland is home to one of the West's largest trade book publishing houses, Graphic Arts Center Publishing.

Oregon has one of the largest salmon-fishing industries in the world, although ocean fisheries have reduced the river fisheries in recent years. Tourism is also strong in the state; Oregon's evergreen mountain forests, waterfalls, pristine lakes (including Crater Lake National Park), and scenic beaches draw visitors year round. The Oregon Shakespeare Festival, held in Ashland, is a tourist draw which complements the southern region of the state's scenic beauty and opportunity for outdoor activities.

Oregon is home to many breweries and Portland has the largest number of breweries of any city in the world.[61]

Portland reportedly has more strip clubs per capita than Las Vegas or San Francisco.[62]

Oregon's gross state product is $132.66 billion as of 2006, making it the 27th largest GSP in the nation.[63]

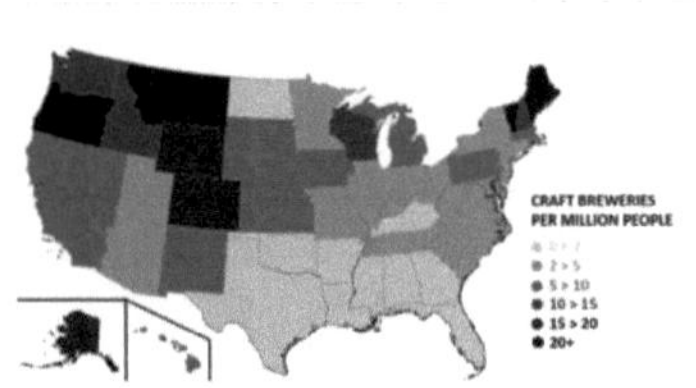

Oregon ranks 4th nationally in craft breweries per capita.

Employment

As of April 2011, the state's unemployment rate is 9.5%.[64] Oregon's largest for-profit employer is Intel, located in the Silicon Forest area on Portland's west side. Intel was the largest employer in Oregon until 2008. As of January 2009, the largest employer in Oregon is Providence Health & Services a non-profit.[65]

Taxes and budgets

Oregon's biennial state budget, $42.4 billion as of 2007, comprises General Funds, Federal Funds, Lottery Funds, and Other Funds. Personal income taxes account for 88% of the General Fund's projected funds.[66] The Lottery Fund, which has grown steadily since the lottery was approved in 1984, exceeded expectations in the 2007 fiscal years, at $604 million.[67]

Oregon is one of only five states that have no sales tax.[68] Oregon voters have been resolute in their opposition to a sales tax, voting proposals down each of the nine times they have been presented.[69] The last vote, for 1993's Measure 1, was defeated by a 72–24% margin.[70]

The state also has a minimum corporate tax of only $10 a year, amounting to 5.6% of the General Fund in the 2005–2007 biennium; data about which businesses pay the minimum is not available to the public.[71] As a result, the state relies on property and income taxes for its revenue. Oregon has the fifth highest personal income tax in the nation. According to the U.S. Census Bureau, Oregon ranked 41st out of the 50 states in taxes per capita in 2005.[72] The average paid of $1,791.45 is higher than only nine other states.[72]

Some local governments levy sales taxes on services: the city of Ashland, for example, collects a 5% sales tax on prepared food.[73]

Oregon is one of six states with a revenue limit.[74] The "kicker law" stipulates that when income tax collections exceed state economists' estimates by 2% or more, any excess must be returned to taxpayers.[75] Since the enactment of the law in 1979, refunds have been issued for seven of the eleven biennia.[76] In 2000, Ballot Measure 86 converted the "kicker" law from statute to the Oregon Constitution, and changed some of its provisions.

Federal payments to county governments, which were granted to replace timber revenue when logging in National Forests was restricted in the 1990s, have been under threat of suspension for several years. This issue dominates the future revenue of rural counties, which have come to rely on the payments in providing essential services.[77]

55 percent of state revenues are spent on public education, 23% on human services (child protective services, Medicaid, and senior services), 17% on public safety, and 5% on other services.[78]

Demographics

Population

Historical populations		
Census	Pop.	%±
1850	12093	—
1860	52465	333.8%
1870	90923	73.3%
1880	174768	92.2%
1890	317704	81.8%
1900	413536	30.2%
1910	672765	62.7%
1920	783389	16.4%
1930	953786	21.8%
1940	1089684	14.2%
1950	1521341	39.6%
1960	1768687	16.3%
1970	2091533	18.3%
1980	2633156	25.9%
1990	2842321	7.9%
2000	3421399	20.4%
2010	3831074	12.0%
U.S. Census Bureau[79]		

As of the census of 2010,[81] Oregon has a population of 3,831,074, which is an increase of 409,675, or 12%, since the year 2000. The population density is 39.9 persons per square mile. There are 1,675,562 housing units, a 15.3% increase over 2000. Among them, 90.7% are occupied.

Hispanics or Latinos make up 11.7% of the total population. Among those who aren't Hispanic or Latino, 78.5% is "white alone," 1.7% is "black or African American alone," 1.1% is "American Indian or Alaska native alone," 3.6% is "Asian alone," 0.3% is

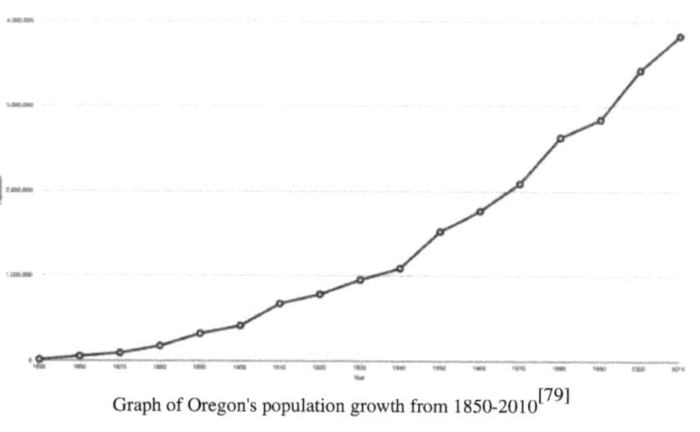

Graph of Oregon's population growth from 1850-2010[79]

"Native Hawaiian and other Pacific Islander alone," 0.1% is "another race alone," and 2.9% is multiracial.

Of the state's total population, 22.6% was under age 18, and 77.4% were 18 or older.

The center of population of Oregon is located in Linn County, in the city of Lyons.[82] More than 57% of the state's population lives in the Portland metropolitan area.[83]

As of 2004, Oregon's population included 309,700 foreign-born residents (accounting for 8.7% of the state population).

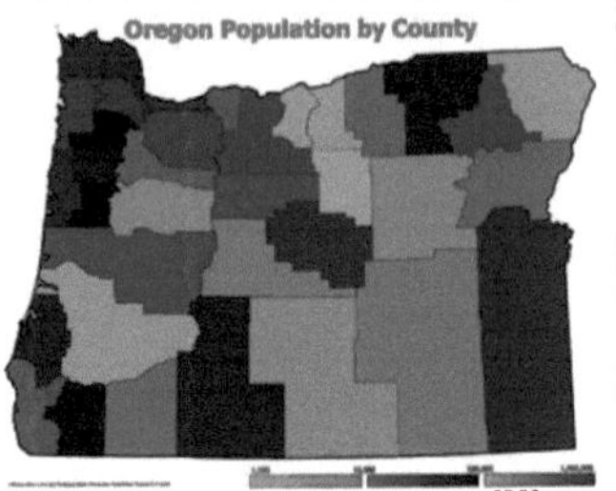

Source: Population Research Center[80]

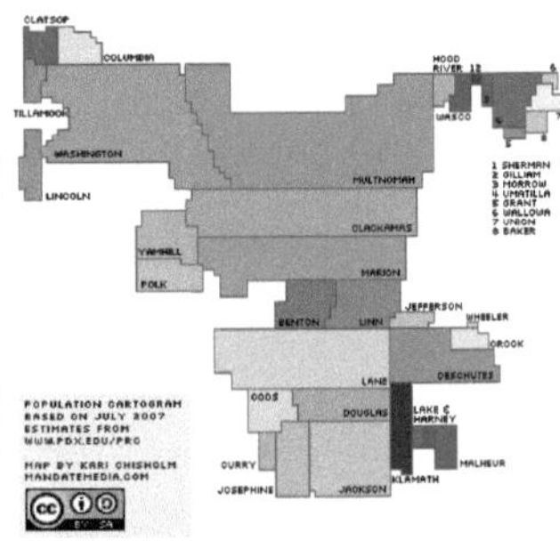

County population cartogram of Oregon.

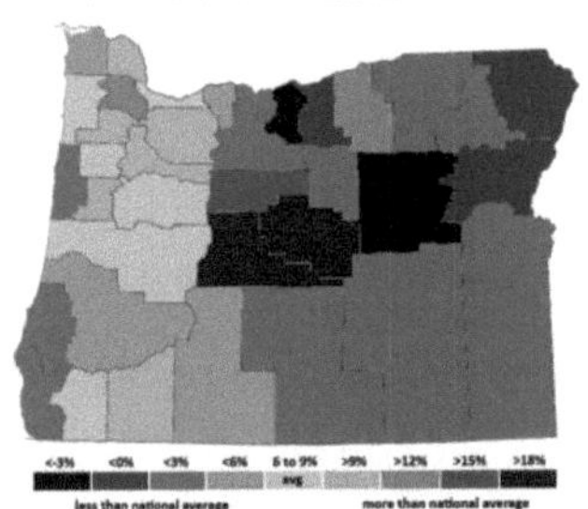

Population Growth by County, 2000–2007.
Green counties grew faster than the national average, while purple counties grew more slowly or, in a few cases, lost population.

Demographics of Oregon (csv) [84]

By race	White	Black	AIAN*	Asian	NHPI*
2000 (total population)	93.45%	2.17%	2.54%	3.75%	0.48%
2000 (Hispanic only)	7.63%	0.17%	0.32%	0.10%	0.05%
2005 (total population)	92.95%	2.38%	2.44%	4.25%	0.50%
2005 (Hispanic only)	9.38%	0.24%	0.34%	0.11%	0.05%
Growth 2000–05 (total population)	5.85%	16.64%	2.45%	20.78%	10.87%
Growth 2000–05 (non-Hispanic only)	3.63%	13.63%	0.62%	20.75%	10.26%
Growth 2000–05 (Hispanic only)	30.84%	52.63%	15.25%	21.84%	16.42%
* AIAN is American Indian or Alaskan Native; NHPI is Native Hawaiian or Pacific Islander					

The largest ancestry groups in the state are:[85]

- 22.5% German
- 14.0% English
- 13.2% Irish
- 8.4% Scandinavian: (1.2% Danish, 3.1% Swedish, & 4.1% Norwegian American)
- 5.0% American
- 3.9% French
- 3.7% Italian
- 3.6% Scottish
- 2.7% Scots-Irish
- 2.6% Dutch
- 1.9% Polish
- 1.4% Russian
- 1.1% Welsh

The largest reported ancestry groups in Oregon are: German (22.5%), English (14.0%), Irish (13.2%), Scandinavian (8.4%) and American (5.0%). Approximately 62% of Oregon residents are wholly or partly of English, Welsh, Irish or Scottish ancestry. Most Oregon counties are inhabited principally by residents of Northwestern-European ancestry. Concentrations of Mexican-Americans are highest in Malheur and Jefferson counties.

The majority of the diversity in Oregon is in the Portland metropolitan area.

Future projections

Projections from the U.S. Census Bureau show Oregon's population increasing to 4,833,918 by 2030, an increase of 41.3% compared to the state's population of 3,421,399 in 2000.[86] The state's own projections forecast a total population of 5,425,408 in 2040.[87]

Religious and secular communities

Major religious affiliations of the people of Oregon are:[88]

- Christian — 67%
  - Protestant — 47%
    - Evangelical — 30%
    - Mainline— 16%
    - Other Protestant — 1%
  - Roman Catholic — 14%

- • Latter Day Saint / Mormon — 5%
- • Other Christian traditions — 1%
- Unaffiliated — 27%
- Buddhist — 2%
- Jewish — 1%
- Muslim — 0.5%
- Other Religions — 2%

The largest denominations by number of adherents in 2000 were the Roman Catholic Church with 348,239; the Church of Jesus Christ of Latter-day Saints with 104,312 (144,808 year-end 2007); and the Assemblies of God with 49,357.[89]

In a 2009 Gallup poll, 69% of Oregonians identified themselves as being Christian.[90] Most of the remainder of the population had no religious affiliation; the 2008 American Religious Identification Survey (ARIS) placed Oregon as tied with Nevada in fifth place of U.S. states having the highest percentage of residents identifying themselves as "non-religious", at 24 percent.[91] [92] Secular organizations include the Center for Inquiry (CFI), the Humanists of Greater Portland (HGP), and the United States Atheists (USA).

During much of the 1990s a group of conservative Christians formed the Oregon Citizens Alliance, and unsuccessfully tried to pass legislation to prevent "gay sensitivity training" in public schools and legal benefits for homosexual couples.[93]

Oregon also contains the largest community of Russian Old Believers to be found in the United States.[94] The Northwest Tibetan Cultural Association is headquartered in Portland, and the New Age film *What the Bleep Do We Know!?* was filmed and had its premiere in Portland. There are an estimated 6,000 to 10,000 Muslims in Oregon.[95]

Education

Primary and secondary

As of 2005, the state had 559,215 students in public primary and secondary schools.[96] There were 199 public school districts at that time, served by 20 education service districts.[96] The five largest school districts as of 2007 were: Portland Public Schools (46,262 students), Salem-Keizer School District (40,106), Beaverton School District (37,821), Hillsboro School District (20,401), and Eugene School District (18,025).[97]

Colleges and universities

Public

The Oregon University System supports seven public universities and one affiliate in the state. The University of Oregon in Eugene is Oregon's flagship liberal arts institution,[98] and was the state's only nationally ranked university by *U.S. News & World Report*.[99] Oregon State University is located in Corvallis and holds the distinction of being the state's flagship research university with top ranked programs

The Memorial Union at OSU

in science, engineering, and agriculture. The university is also the state's highest ranking university/college in a world survey of academic merit.[100]

The state's urban Portland State University has Oregon's largest enrollment. The state has three regional universities: Western Oregon University in Monmouth, Southern Oregon University in Ashland, and Eastern Oregon University in La Grande. The Oregon Institute of Technology has its campus in Klamath Falls. The affiliate Oregon Health & Science University (OHSU) comprises a medical, dental, and nursing school in Portland and a science and engineering school in Hillsboro. It rated 2nd among US best medical schools based on research by The Med School 100.[101]

Oregon has historically struggled to fund higher education. Recently, Oregon has cut its higher education budget over 2002–2006 and now Oregon ranks 46th in the country in state spending per student. However, 2007 legislation forced tuition increases to cap at 3% per year, and funded the OUS far beyond the requested governor's budget.[102]

The state also supports 17 community colleges.

Private

Oregon is home to a wide variety of private colleges. The University of Portland and Marylhurst University are Catholic institutions in the Portland area. Reed College, Concordia University, Lewis & Clark College, Multnomah Bible College, Portland Bible College, Warner Pacific College, Cascade College, the National College of Natural Medicine and Western Seminary, a theological graduate school, are also in Portland. Pacific University is in the Portland suburb of Forest Grove.

There are also private colleges further south in the Willamette Valley. McMinnville has Linfield College, while nearby Newberg is home to George Fox University. Salem is home to two private schools, Willamette University (the state's oldest, established during the provisional period) and Corban University. Also located near Salem is Mount Angel Seminary, one of America's largest Roman Catholic seminaries. Eugene is home to three private colleges: Northwest Christian University, Eugene Bible College, and Gutenberg College.

Sports

Oregon is home to two professional sports teams which are both based in Portland: Portland Trail Blazers of the NBA and the Portland Timbers of MLS.[103]

Until 2011, the only major professional sports team in Oregon was the Portland Trail Blazers of the National Basketball Association. From the 1970s to the 1990s, the Blazers were one of the most successful teams in the NBA in terms of both win-loss record and attendance. In the early 21st century, the team's popularity declined due to personnel and financial issues, but revived after the departure of controversial players and the acquisition of new players such as Brandon Roy, LaMarcus Aldridge, and Greg Oden.[104] [105]

The Rose Garden during a Portland Trail Blazers game

The Blazers play in the Rose Garden in Portland's Lloyd District, which is also home to the Portland Winterhawks of the junior-league Western Hockey League.[106]

The Timbers play at Jeld-Wen Field, which is just west of downtown Portland. The Timbers repurposed the formerly multi-use stadium into a soccer-only configuration in fall 2010, increasing the seating in the process.[107]

Portland has had minor league baseball teams in the past, including the Portland Beavers and Portland Rockies, who played most recently at PGE Park. Portland has also actively pursued a Major League Baseball team.[108] It was announced in March 2009 that the Portland Timbers will begin MLS play starting in 2011.[103]

Eugene and Salem also have minor-league baseball teams. The Eugene Emeralds and the Salem-Keizer Volcanoes both play in the Single-A Northwest League.[109] Oregon also has four teams in the fledgling International Basketball League: the Portland Chinooks, Central Oregon Hotshots, Salem Stampede, and the Eugene Chargers.[110]

The Oregon State Beavers and the University of Oregon Ducks football teams of the Pacific-12 Conference meet annually in the Civil War. Both schools have had recent success in other sports as well: Oregon State won back-to-back college baseball championships in 2006 and 2007,[111] and the University of Oregon won back-to-back NCAA men's cross country championships in 2007 and 2008.[112]

Sister states

- ■ People's Republic of China, Fujian Province - 1984[113]
- ■ Republic of China, Taiwan - 1985[113]
- ● Japan, Toyama Prefecture - 1991[113] [114]
- ⦂●⦂ Republic of Korea, Jeollanam-do Province - 1996[113] [114]
- ▬ Iraq, Iraqi Kurdistan - 2005[115]

See also

- Outline of Oregon
- Index of Oregon-related articles
- List of people from Oregon
- National Register of Historic Places listings in Oregon

References

[1]   Calvin Hall (2007-01-30). "English as Oregon's official language? It could happen" (http://media.www.dailyemerald.com/media/storage/paper859/news/2007/01/30/News/English.As.Oregons.Official.Language.It.Could.Happen-2685082.shtml). *The Oregon Daily Emerald*. . Retrieved 2007-05-08.

[2]   "2010 Resident Population Data: Population Change" (http://2010.census.gov/2010census/data/apportionment-pop-text.php). US Census. . Retrieved December 21, 2010.

[3]   "2010 Resident Population Data: Population Density" (http://2010.census.gov/2010census/data/apportionment-dens-text.php). US Census. . Retrieved December 21, 2010.

[4]   "Mount Hood Highest Point" (http://www.ngs.noaa.gov/cgi-bin/ds_mark.prl?PidBox=RC2244). *NGS data sheet*. U.S. National Geodetic Survey. . Retrieved October 24, 2011.

[5]   "Elevations and Distances in the United States" (http://egsc.usgs.gov/isb/pubs/booklets/elvadist/elvadist.html). United States Geological Survey. 2001. . Retrieved October 24, 2011.

[6]   Elevation adjusted to North American Vertical Datum of 1988.

[7]   Map of Oregon. (http://nationalatlas.gov/printable/images/pdf/reference/pagegen_or.pdf) Reston, Virginia: Interior Geological Survey, 2004

[8]   Oregon Fast Facts. (http://traveloregon.mediaroom.com/index.php?s=23&item=33) Travel Oregon.

[9]   "U.S. Census Bureau — State & County QuickFacts — Oregon" (http://2010.census.gov/2010census/data/index.php). . Retrieved 2010-12-21.

[10]   Robbins, William G. (2005). *Oregon: This Storied Land*. Oregon Historical Society Press. ISBN 0987595-286-0.

[11]   "Oregon History: Great Basin" (http://bluebook.state.or.us/cultural/history/history04.htm). *Oregon Blue Book*. Oregon State Archives. . Retrieved 2007-09-02.

[12]   "Oregon History: Northwest Coast" (http://bluebook.state.or.us/cultural/history/history02.htm). *Oregon Blue Book*. Oregon State Archives. . Retrieved 2007-09-02.

[13]   "Confederated Tribes of the Grand Ronde: Culture" (http://www.grandronde.org/culture/). . Retrieved 2007-09-02.

[14]   "Oregon History: Columbia Plateau" (http://bluebook.state.or.us/cultural/history/history03.htm). *Oregon Blue Book*. Oregon State Archives. . Retrieved 2007-09-02.

[15]   "Chronology of events, 1543-1859" (http://arcweb.sos.state.or.us/echoes/chronology.html). *Echoes of Oregon History Learning Guide*. Oregon State Archives. . Retrieved 2011-08-11.

[16]   Loy, Willam G.; Stuart Allan, Aileen R. Buckley, James E. Meacham (2001). *Atlas of Oregon*. University of Oregon Press. pp. 12–13. ISBN 0-87114-101-9.

[17]   McLagan, Elizabeth (1980). *A Peculiar Paradise*. Georgian Press. ISBN 0-96034-082-3.

[18]   Oregon Almanac (http://www.bluebook.state.or.us/facts/almanac/almanac04.htm)

[19]   Where does the name "Oregon" come from? (http://www.sos.state.or.us/bbook/misc/about/faq.htm#oregon) from the online edition of the Oregon Blue Book

[20]   Elliott, T.C. (June 1921). "The Origin of the Name Oregon" (http://books.google.com/books?id=P-oXAAAAYAAJ&pg=PA99). *Oregon Historical Quarterly* **XXIII** (2): 99–100. ISSN 0030-4727. OCLC 1714620. . Retrieved 11 October 2010.

[21] Miller, Joaquin (1904). "Sea of Silence", *Sunset*, 396(13):5. (http://books.google.com/books?id=GWhpzV9XeQUC&dq=sea of silence miller&pg=PA395#v=onepage&q=sea of silence miller&f=false)

[22] "Oregon" (http://www.merriam-webster.com/dictionary/oregon). Merriam-Webster Online Dictionary. . Retrieved 2006-09-14.

[23] Banks, Don (21 April 2002). "Harrington confident about Detroit QB challenge." (http://sportsillustrated.cnn.com/inside_game/don_banks/news/2002/04/21/harrington_intro/) *Sports Illustrated.*

[24] Bellamy, Ron (6 October 2003). "See no evil, hear no evil" (http://news.google.com/newspapers?id=MFFWAAAAIBAJ&sjid=qusDAAAAIBAJ&dq=Joey Harrington scoffs at criticism as he struggles to right the Lions&pg=3329,1524986). *The Register-Guard.* . Retrieved 1 June 2011.

[25] "Yellow/Green ORYGUN Block Letter Outside Decal" (http://spiritduck.uoduckstore.com/Yellow_Green_ORYGUN_Block_Letter_Outside_Decal_p/76386407024.htm). UO Duck Store. . Retrieved 2011-08-03.

[26] United States—States; and Puerto Rico: GCT-PH1-R. Population, Housing Units, Area, and Density. (http://factfinder.census.gov/servlet/GCTTable?_bm=n&_lang=en&mt_name=DEC_2000_SF1_U_GCTPH1R_US9S&format=US-9S&_box_head_nbr=GCT-PH1-R&ds_name=DEC_2000_SF1_U&geo_id=01000US) U.S. Census Bureau. Retrieved on March 28, 2008.

[27] "Elevations and Distances in the United States" (http://egsc.usgs.gov/isb/pubs/booklets/elvadist/elvadist.html). U.S Geological Survey. 29 April 2005. . Retrieved November 7, 2006.

[28] "Crater Lake National Park" (http://www.nps.gov/crla/index.htm). U.S. National Park Service. . Retrieved 2006-11-22.

[29] "D River State Recreation Site" (http://www.oregonstateparks.org/park_214.php). *Oregon Parks and Recreation Department.* . Retrieved 2007-05-11.

[30] "World's Shortest River" (http://montanakids.com/db_engine/presentations/presentation.asp?pid=192). *Travel Montana.* . Retrieved 2007-05-11.

[31] "Mill Ends Park" (http://www.portlandonline.com/parks/finder/index.cfm?PropertyID=265&action=ViewPark). *Portland Parks and Recreation.* . Retrieved 2007-05-11.

[32] Beale, Bob (10 April 2003). "Humungous fungus: world's largest organism?" (http://www.abc.net.au/science/news/enviro/EnviroRepublish_828525.htm) Environment & Nature News, ABC Online. Accessed January 2, 2007.

[33] Western States Data Public Land Acreage (http://www.wildlandfire.com/docs/2007/western-states-data-public-land.htm) (13 November 2007).

[34] "2010 Census Redistricting Data" (http://factfinder2.census.gov/faces/tableservices/jsf/pages/productview.xhtml?pid=DEC_10_PL_GCTPL2.ST13&prodType=table). US Census Bureau. . Retrieved 2011-03-15.

[35] 50 Fastest-Growing Metro Areas Concentrated in West and South. (http://www.census.gov/newsroom/releases/archives/population/cb07-51.html) U.S. Census Bureau 2005. Retrieved October 16, 2007.

[36] Allen, Cain (2006). "The Oregon History Project- A Pacific Republic" (http://www.ohs.org/education/oregonhistory/historical_records/dspDocument.cfm?doc_ID=92F96702-9FCF-BAC0-3411BF129F1F23DA). Oregon Historical Society. Retrieved 2010-10-08.

[37] Oregon Secretary of State. "A Brief History of the Oregon Territorial Period" (http://arcweb.sos.state.or.us/echoes/history.html). State of Oregon. . Retrieved 2006-08-09.

[38] "Constitution of Oregon (Article V)" (http://bluebook.state.or.us/state/constitution/constitution05.htm). *Oregon Blue Book.* State of Oregon. 2007. . Retrieved 2008-03-12.

[39] ORS 653.025 (http://www.leg.state.or.us/ors/653.html).

[40] "November 2, 2004, General Election Abstract of Votes: STATE MEASURE NO. 36" (http://www.sos.state.or.us/elections/nov22004/abstract/m36.pdf). Oregon Secretary of State. . Retrieved 2008-11-17.

[41] Bradbury, Bill (November 6, 2007). "Official Results – November 6, 2007 Special Election" (http://www.sos.state.or.us/elections/nov62007/). *Elections Division.* Oregon Secretary of State. . Retrieved December 27, 2008.

[42] "November 7, 2006, general election abstracts of votes: state measure no. 39" (http://www.sos.state.or.us/elections/nov72006/results/m39.pdf). State of Oregon. . Retrieved 12 March 2011.

[43] S. Spacek, 2011 American State Litter Scorecard: New Rankings for An Increasingly Environmetally-Concerned Populace.

[44] "United States Bankruptcy Court, District of Oregon" (http://www.orb.uscourts.gov/). U.S. Courts. . Retrieved 2008-12-14.

[45] Leip, David. "2008 presidential general election results" (http://uselectionatlas.org/RESULTS/). Retrieved 2010-10-12.

[46] Silver, Nate (May 17, 2008). "Oregon: Swing state or latte-drinking, Prius-driving lesbian commune?" (http://www.fivethirtyeight.com/2008/05/oregon-swing-state-or-latte-drinking.html). FiveThirtyEight.com. .

[47] "State Initiative and Referendum Summary" (http://www.iandrinstitute.org/statewide_i&r.htm). State Initiative & Referendum Institute at USC. . Retrieved 2006-11-27.

[48] "Eighth Annual Report on Oregon's Death with Dignity Act" (http://www.oregon.gov/DHS/ph/pas/docs/year8.pdf) (PDF). Oregon Department of Human Services. March 9, 2006. . Retrieved 2007-06-11.

[49] "Voting In Oregon - Vote By Mail." (http://www.co.multnomah.or.us/dbcs/elections/election_information/voting_in_oregon.shtml) Multnomah County, Oregon.

[50] (http://www.usgovernmentrevenue.com/state_rev_summary.php?chart=Z0&year=2010&units=d&rank=a) Bureau of Economic Analysis.

[51] McNab, W. Henry; Avers, Peter E (July 1994). *Ecological Subregions of the United States.* Chapter 24. (http://www.fs.fed.us/land/pubs/ecoregions/) U.S. Forest Service and Dept. of Agriculture.

[52] "Industry Facts" (http://oregonwine.org/press/StateWineFacts2005.pdf) (PDF). Oregon Winegrowers Association. . Retrieved 2006-11-23.

[53] "Oregon Forest Facts: 25-Year Harvest History" (http://www.oregonforests.org/factbook/Harvest_History(24).html). Oregon Forest Resources Institute. . Retrieved 2007-03-07.

[54] "Forest Economics and Employment" (http://www.oregonforests.org/factbook/economics(29_30).html). Oregon Forest Resources Institute. . Retrieved 2007-03-08.

[55] "Oregon economy" (http://www.e-referencedesk.com/resources/state-economy/oregon.html). e-ReferenceDesk. Retrieved 2010-11-05.

[56] Don Hamilton (2002-07-19). "Matt Groening's Portland" (http://portlandtribune.com/news/story.php?story_id=12392). *The Portland Tribune*. . Retrieved 2007-03-07.

[57] http://beargrylls.blogspot.com/2008/11/man-vs-wildborn-survivor-complete.html#links

[58] "Bright spots amid the turmoil" (http://www.oregonlive.com/business/oregonian/index.ssf?/base/business/1199161505105830.xml& coll=7). *The Oregonian*. January 1, 2008. p. D3. . Retrieved 2007-01-01.

[59] Rogoway, Mike (January 15, 2009). "Intel profits slide, company uncertain about outlook" (http://www.oregonlive.com/business/index. ssf/2009/01/intel_profits_slide_company_un.html). *The Oregonian*. . Retrieved 2009-01-16.

[60] "Genentech Selects Hillsboro" (http://www.hillchamber.org/memberservices/in_the_news.asp#Genentech). Hillsboro Chamber of Commerce. . Retrieved 2007-03-21.

[61] "Oregon's Beer Week gets under way." (http://goliath.ecnext.com/coms2/gi_0198-242714/Oregon-s-Beer-Week-gets.html). Knight-Ridder Tribune News Service. 2005-07-05. . Retrieved 2007-10-22.

[62] Moore, Adam S.; Beck, Byron (November 8, 2004). "Bump and grind" (http://www.wweek.com/story.php?story=6093). *Willamette Week*. . Retrieved 2007-02-01.

[63] "Gross Domestic Product (GDP) by State, 2006" (http://www.bea.gov/newsreleases/regional/gdp_state/gsp_newsrelease.htm). Bureau of Economic Analysis — U.S. Department of Commerce. . Retrieved 2007-06-10.

[64] http://www.google.com/publicdata?ds=usunemployment&met_y=unemployment_rate&idim=state:ST410000&dl=en&hl=en& q=oregon+unemployment+rate

[65] Rogoway, Mike. "Oregon's largest private employer" (http://blog.oregonlive.com/siliconforest/2009/01/ oregons_largest_private_employ.html). The Oregonian. . Retrieved 2011-01-25.

[66] "Government Finance: State Government" (http://bluebook.state.or.us/state/govtfinance/govtfinance01.htm). *Oregon Blue Book*. . Retrieved 2007-06-20.

[67] Har, Janie (2007-06-20). "Your loss is state's record game" (http://0-docs.newsbank.com.catalog.multcolib.org/openurl?ctx_ver=z39. 88-2004&rft_id=info:sid/iw.newsbank.com:NewsBank:ORGB&rft_val_format=info:ofi/fmt:kev:mtx:ctx& rft_dat=119E94A840D2FBB8&svc_dat=InfoWeb:aggregated4&req_dat=0D10F2CADB4B24C0). *The Oregonian*. . Retrieved 2007-06-20.

[68] "State Sales Tax Rates" (http://www.taxadmin.org/FTA/rate/sales.html). Federation of Tax Administrators. 2008-01-01. . Retrieved 2008-04-02.

[69] "25th Anniversary Issue: 1993" (http://www.wweek.com/html/25-1993.html). *Willamette Week*. . Retrieved 2007-06-11.

[70] "Initiative, Referendum and Recall: 1988–1995" (http://bluebook.state.or.us/state/elections/elections21.htm). *Oregon Blue Book*. State of Oregon. . Retrieved 2007-06-11.

[71] Sheketoff, Charles (2007-03-27). "As Maryland Goes, So Should Oregon" (http://salem-news.com/articles/march272007/ oregon_mrlnd_32707.php). *Salem News*. . Retrieved 2007-06-10.

[72] "Oregon ranks 41st in taxes per capita" (http://www.bizjournals.com/portland/stories/2006/03/27/daily28.html). Portland Business Journal. 2006-03-31. . Retrieved 2007-06-10.

[73] "Food and Beverage Tax" (http://www.ashland.or.us/Page.asp?NavID=9180). City of Ashland. . Retrieved 2007-06-10.

[74] "Oregon's 2% Kicker" (http://www.leg.state.or.us/comm/lro/rr02-07.pdf) (PDF). *Oregon State Leglislative Review Office*. . Retrieved 2007-06-10.

[75] Cain, Brad (March 2, 2006). "Kicker tax rebate eyed to help school and state budgets" (http://www.katu.com/news/3617476.html). KATU.com. . Retrieved 2006-06-10.

[76] "2 Percent Surplus Refund (Kicker) History" (http://www.oregon.gov/DOR/NEWS/docs/kicker.pdf) (PDF). State of Oregon. . Retrieved 2007-06-10.

[77] Cooper, Matt (2007-03-09). "County may scrub income tax" (http://www2.registerguard.com/cms/index.php/static/search/archive/ ?q=County+may+scrub+income+tax). *The Register-Guard*. . Retrieved 2007-03-09.

[78] "2006 Oregon full-year resident tax form instructions." (http://www.oregon.gov/DOR/PERTAX/docs/2006Forms/101-043-06.pdf) Oregon.Gov.

[79] "Oregon" (http://www.census.gov/dmd/www/resapport/states/oregon.pdf). *Resident Population and Apportionment of the U.S. House of Representatives*. U.S. Census Bureau. December 27, 2000. . Retrieved 2009-08-28.

[80] "Annual Population Estimates" (http://www.pdx.edu/prc/annualorpopulation.html). Portland State University Population Research Center. . Retrieved 2008-03-03.

[81] "2010 census profiles: Oregon and its counties" (http://www.pdx.edu/sites/www.pdx.edu.prc/files/media_assets/ 2010_PL94_counties_updated.pdf). Portland State University Population Research Center. . Retrieved 15 May 2011.

[82] "Population and Population Centers by State: 2000" (http://www.census.gov/geo/www/cenpop/statecenters.txt). U.S. Census Bureau. . Retrieved 2006-11-23.

[83] "JULY 1, 2006 Population estimates for Metropolitan Combined Statistical Areas" (http://web.archive.org/web/20070920212740/http://
      www.census.gov/population/www/estimates/metro_general/2006/CBSA-EST2006-01.csv) (csv). U.S. Census Bureau. Archived from
      the original (http://www.census.gov/population/www/estimates/metro_general/2006/CBSA-EST2006-01.csv) on 2007-09-20. .
      Retrieved 2007-10-19.

[84] http://www.census.gov/popest/states/asrh/tables/SC-EST2005-03-41.csv

[85] "2006-2008 American Community Survey 3-year estimates." (http://factfinder.census.gov/servlet/ADPTable?_bm=y&-context=adp&
      -qr_name=ACS_2008_3YR_G00_DP3YR2&-ds_name=ACS_2008_3YR_G00_&-tree_id=3308&-redoLog=false&-_caller=geoselect&
      -geo_id=04000US41&-format=&-_lang=en) U.S. Census Bureau.

[86] United States Census Bureau (2005-04-21). "Interim Projections of the Total Population for the United States and States: April 1, 2000 to
      July 1, 2030" (http://www.census.gov/population/projections/SummaryTabA1.pdf). . Retrieved 2010-08-18.

[87] Office of Economic Analysis (April 2004). "State and County Population Forecasts and Components of Change, 2000 to 2040" (http://
      www.oregon.gov/DAS/OEA/docs/demographic/pop_components.xls). Oregon Department of Administrative Services. . Retrieved
      2010-08-25.

[88] "U.S. Religious Landscape Survey." (http://religions.pewforum.org/maps) The Pew Forum on Religion & Public Life. Retrieved
      2010-02-05.

[89] "State membership report- Oregon." (http://www.thearda.com/mapsReports/reports/state/41_2000.asp) The Association of Religion
      Data Archives.

[90] Newport, Frank (7 August 2009). "Religious identity: States differ widely." (http://www.gallup.com/poll/122075/
      Religious-Identity-States-Differ-Widely.aspx#2) Gallup. Retrieved 2009-12-23.

[91] Kosmin, Barry A.; Keysar, Ariela. "American Religious Identification Survey." (http://www.americanreligionsurvey-aris.org/reports/
      ARIS_Report_2008.pdf) Hartford: Trinity College. Retrieved 2009-12-23.

[92] Kosmin, Barry A.; Keysar, Ariela with Cragun, Ryan; Navarro-Rivera, Juhem. "American nones: The profile of the no religion population."
      (http://www.americanreligionsurvey-aris.org/reports/NONES_08.pdf) Hartford: Trinity College. Retrieved 2009-12-23.

[93] Wentz, Patty (11 February 1998). "He's back." (http://wweek.com/html/cover021198.html) *Willamette Week*. Retrieved on March 14,
      2008.

[94] Binus, Joshua. "The Oregon History Project: Russian Old Believers." (http://www.ohs.org/education/oregonhistory/historical_records/
      dspDocument.cfm?doc_ID=764E6BED-FFC4-C034-9A5563F41CE37080) Oregon Historical Society. Retrieved on March 14, 2008.

[95] Islam in Oregon and America—The Facts (http://www.metpdx.org/resources/) (dead link)

[96] "Oregon Blue Book: Oregon Almanac: Native Americans to shoes, oldest." (http://bluebook.state.or.us/facts/almanac/almanac05)
      Oregon Secretary of State. Retrieved on March 28, 2008.

[97] "Oregon public school enrollment increases during 2007-08." (http://www.ode.state.or.us/news/releases/?yr=0000&kw=&rid=610)
      Oregon Department of Education. Retrieved on March 28, 2008.

[98] Wood, Shelby Oppel (2006-05-01). "UO weighs new diversity plan amid simmering racial tensions" (http://www.oregonlive.com/search/
      oregonian/). *The Oregonian*. .

[99] USNews.com: America's Best Colleges 2008: National Universities: Best Schools (http://colleges.usnews.rankingsandreviews.com/
      usnews/edu/college/rankings/brief/t1natudoc_brief.php)

[100] "Top 500 World Universities" (http://ed.sjtu.edu.cn/ranking.htm). . Retrieved 2007-10-18. (dead link)

[101] Good University Ranking Guide (http://whichuniversitybest.blogspot.com/2010/03/ohsu-school-of-medicine.html)

[102] "Higher education get higher priority." (http://media.www.dailyemerald.com/media/storage/paper859/news/2007/06/29/News/
      Higher.Education.Gets.Higher.Priority-2919794.shtml) *The Oregon Daily Emerald*, 29 June 2007. Retrieved 2007-07-08.

[103] "MLS awards team to Portland for 2011." (http://portlandtimbers.com/newsroom/headlines/index.html?article_id=1108) Portland
      Timbers, 20 March 2009.

[104] Smith, Sam (October 18, 2006). "Blazers stalled until bad apples go" (http://www.msnbc.msn.com/id/15321476/). MSNBC.com. .
      Retrieved 2008-01-15.

[105] Mejia, Tony (October 13, 2007). "Oden's loss hurts, but team in good hands" (http://www.cbsnews.com/stories/sportsline/
      main10406427.shtml). CBSNews.com. . Retrieved 2008-01-15.

[106] "Rose Quarter Venues" (http://www.rosequarter.com/RoseQuarter/Venues/tabid/84/Default.aspx). RoseQuarter.com. . Retrieved
      2008-01-15.

[107] "PGE Park Teams and Events" (http://www.pgepark.com/stadium/events/). PGEPark.com. . Retrieved 2008-01-15.

[108] "Oregon Stadium Campaign" (http://www.oregonstadiumcampaign.com/). Oregon Stadium Campaign. . Retrieved 2008-01-14.

[109] "Northwest League." (http://web.minorleaguebaseball.com/index.jsp?sid=l126) Minor League Baseball. Retrieved 2008-01-15.

[110] "International Basketball League." (http://www.iblhoopsonline.com/) International Basketball League. Retrieved 2008-01-15.

[111] Beseda, Jim (12 August 2010). "Oregon State baseball: Coach Pat Casey praises ex-Beaver Darwin Barney" (http://blog.oregonlive.com/
      behindbeaversbeat/2010/08/oregon_state_baseball_coach_pa_2.html). *The Oregonian* (Portland, Oregon). Retrieved 2010-10-08.

[112] The Associated Press (8 January 2009). "Oregon men, Washington women win titles" (http://sports.espn.go.com/ncaa/news/
      story?id=3723958). ESPN. Retrieved 2010-10-08.

[113] Van Winkle, Teresa (June 2008). "Background brief on international trade." (http://www.leg.state.or.us/comm/commsrvs/
      background_briefs2008/briefs/EconomyBusinessLabor/InternationalTrade.pdf) State of Oregon. Retrieved 2008-07-21.

[114] "Governor's mission to Asia will stress trade and cultural ties." (http://arcweb.sos.state.or.us/governors/Kitzhaber/web_pages/
       governor/press/p951024.htm) Oregon Secretary of State, 24 October 1995. Retrieved 2008-04-02.
[115] http://www.leg.state.or.us/05orlaws/sessresmem.dir/scr0003ses.htm

Further reading

- *Excursion to the Oregon* by John Kirk Townsend (http://www.secstate.wa.gov/history/publications_detail.
  aspx?p=33)
- *New map of Texas, Oregon and California with the regions adjoining, compiled from the more recent authorities*
  by Samuel Augustus Mitchell (http://www.secstate.wa.gov/history/maps_detail.aspx?m=14)
- *Accompaniment to Mitchell's New map of Texas, Oregon, and California, with the regions adjoining* by Samuel
  Augustus Mitchell (http://www.secstate.wa.gov/history/publications_detail.aspx?p=26)
- O'Hara, E. (1911). Oregon (http://www.newadvent.org/cathen/11288a.htm). In the *Catholic Encyclopedia*.
  New York: Robert Appleton Company. Retrieved July 25, 2009, from New Advent.

External links

- State of Oregon (http://www.oregon.gov/) (official website)
- Oregon Blue Book (http://bluebook.state.or.us/), the online version of the state's official directory and fact
  book
- TravelOregon.com (http://www.TravelOregon.com/) an official website of the Oregon Tourism Commission
- Oregon Historical Society (http://www.ohs.org/)
- Oregon State Databases (http://wikis.ala.org/godort/index.php/Oregon), an annotated list, in wiki form, of
  searchable databases produced by Oregon state agencies and compiled by the Government Documents
  Roundtable of the American Library Association
- Real-time, geographic, and other scientific resources of Oregon (http://www.usgs.gov/state/state.
  asp?State=OR) from the United States Geological Survey
- Oregon Quickfacts (http://quickfacts.census.gov/qfd/states/41000.html) from the United States Census
  Bureau
- Oregon State Facts (http://www.ers.usda.gov/StateFacts/OR.htm) from the United States Department of
  Agriculture
- Oregon (http://www.dmoz.org/Regional/North_America/United_States/Oregon/) at the Open Directory
  Project
- OpenStreetMap has geographic data related to Oregon (http://www.openstreetmap.org/browse/relation/
  165476)

gag:Oregon mrj:Орегон

Blue_Mountains_(Oregon)

The **Blue Mountains** are a mountain range in the western United States, located largely in northeastern Oregon and stretching into southeastern Washington. The range, situated in the Pacific Northwest, has an area of 4060 square miles (10500 km^2), stretching east and southeast of Pendleton, Oregon, to the Snake River along the Oregon-Idaho border.

Geologically, the range is a part of the larger rugged Columbia River Plateau, located in the dry area of Oregon east of the Cascade Range. The highest peaks in the range include the Elkhorn Mountains at 9108 feet (2776 m),[1] Strawberry Mountain at 9038 feet (2755 m), and Mount Ireland at 8304 feet (2531 m).[2] The nearby Wallowa Mountains, east of the main range near the Snake River, are sometimes included as a subrange of the Blue Mountains.

In the middle 19th century, the Blue Mountains were a formidable obstacle on the Oregon Trail and was often the last mountain range American pioneers had to cross before reaching either southeast Washington near Walla Walla or passing down the Columbia River Gorge to end of the Oregon Trail in the Willamette Valley near Oregon City. The range today is traversed by Interstate 84, which crosses the crest of the range at a 4193 feet (1278 m) summit, from south-southeast to north-northwest between La Grande and Pendleton. The community of Baker City sits along the southeastern flank of the range. U.S. Route 26 crosses the southern portion of the range, reaching a summit of 5098 feet (1554 m) at Blue Mountain Pass.

The Blue Mountains in Washington, seen from the west

Much of the range is included in the Malheur National Forest, Umatilla National Forest, and Wallowa–Whitman National Forest. Several wilderness areas encompass remote parts of the range, including the Umatilla Wilderness, the North Fork John Day Wilderness, Strawberry Mountain Wilderness, Monument Rock Wilderness, all of which are in Oregon. The Wenaha–Tucannon Wilderness sits astride the Oregon-Washington border.

The range is drained by several rivers, including the Grand Ronde and Tucannon, tributaries of the Snake, as well as the forks of the John Day, Umatilla and Walla Walla rivers, tributaries of the Columbia.

References

[1] "Oregon's Highest Named Summits" (http://americasroof.com/highest/or.shtml). . Retrieved 2010-03-07.
[2] "Feature Detail Report Mount Ireland" (http://geonames.usgs.gov/pls/gnispublic/f?p=gnispq:3:::NO::P3_FID:1124418). US Geological Survey. . Retrieved 2008-07-25.

External links

- "Blue Mountains (mountain range)" (http://www.getty.edu/vow/TGNFullDisplay?find=Blue+Mountains& place=&nation=&prev_page=1&english=Y&subjectid=1108805). Getty Thesaurus of Geographic Names. 2004. Retrieved 2007-07-28.

Canyon_City,_Oregon

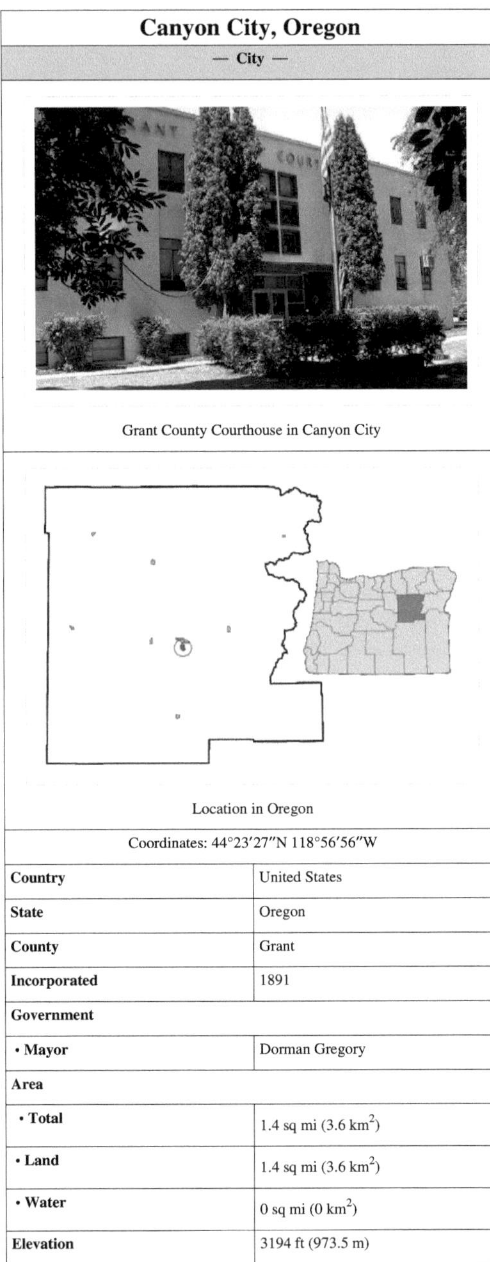

Canyon City, Oregon
— City —

Grant County Courthouse in Canyon City

Location in Oregon

Coordinates: 44°23′27″N 118°56′56″W

Country	United States
State	Oregon
County	Grant
Incorporated	1891
Government	
• Mayor	Dorman Gregory
Area	
• Total	1.4 sq mi (3.6 km^2)
• Land	1.4 sq mi (3.6 km^2)
• Water	0 sq mi (0 km^2)
Elevation	3194 ft (973.5 m)

Population (2010)	
• **Total**	703
• **Density**	430.3/sq mi (158.8/km^2)
Time zone	Pacific (UTC-8)
• **Summer (DST)**	Pacific (UTC-7)
ZIP code	97820
Area code(s)	541
FIPS code	41-10950[1]
GNIS feature ID	1139318[2]

Canyon City is a city in Grant County, Oregon, United States. It is the county seat of Grant County, and is about a mile south of John Day on U.S. Highway 395. As of the 2010 census, the city had a total population of 703.

History

Canyon City was established in June 1862 as a result of the discovery of gold in Canyon Creek, which flows through the town. The discovery led to land between Canyon City and John Day being worth $500 per square yard ($598/m²).[3] Panning for gold could yield several ounces in each pan of sediment. At the peak of the local gold rush, it is claimed that 10,000 people lived in Canyon City, making it larger than Portland at the time.

The preeminent geologist, Waldemar Lindgren, during his 1900 visit to the gold belt of the Blue Mountains, estimated that no more than $16 million in gold had been recovered from Canyon Creek by then. At the fixed value of $20.67 per troy ounce, this would correspond to roughly **unknown operator: u','** troy ounces (**unknown operator: u'strong'unknown operator: u','**kg), worth today (at $1,000 an ounce) about $800-million.

The town was platted in 1862 and was chosen as the county seat two years later, when Grant County was split from Wasco County. It was incorporated in 1891.

Geography

Canyon City is located a few miles northwest of Canyon Mountain, at an altitude of over 3100 feet (940 m).

According to the United States Census Bureau, the city has a total area of 1.4 square miles (3.6 km^2); none of the area is covered with water.[4]

Demographics

As of the census[1] of 2000, there are 669 people in the town, organized into 257 households and 180 families. The population density is 480.3 people per square mile (185.8/km²). There are 294 housing units at an average density of 211.1 per square mile (81.7/km²). The racial makeup of the town is 90.58% White, 2.84% Native American, 0.15% Black or African American, 2.09% from other races, and 4.33% from two or more races. No one identified himself as Asian or Pacific Islander. 4.63% of the population are Hispanic or Latino of any race.

There are 257 households out of which 37.7% have children under the age of 18 living with them, 54.1% are married couples living together, 11.7% have a female householder with no husband present, and 29.6% are non-families. 24.9% of all households are made up of individuals and 8.6% have someone living alone who is 65 years of age or older. The average household size is 2.46 and the average family size is 2.94.

In the town the population is spread out with 28.7% under the age of 18, 6.4% from 18 to 24, 31.2% from 25 to 44, 20.5% from 45 to 64, and 13.2% who are 65 years of age or older. The median age is 35 years. For every 100 females there are 99.7 males. For every 100 females age 18 and over, there are 107.4 males.

The median income for a household in the town is $29,940, and the median income for a family is $41,458. Males have a median income of $31,167 versus $23,438 for females. The per capita income for the town is $14,404. 7.8% of the population and 5.7% of families are below the poverty line. Out of the total population, 9.2% of those under the age of 18 and 0.0% of those 65 and older are living below the poverty line.

Notable people

- Joaquin Miller, poet and essayist[5]

References

[1] "American FactFinder" (http://factfinder.census.gov). United States Census Bureau. . Retrieved 2008-01-31.

[2] "US Board on Geographic Names" (http://geonames.usgs.gov). United States Geological Survey. 2007-10-25. . Retrieved 2008-01-31.

[3] *Oregon's Golden Years* (ISBN 0-87004-254-8)

[4] "US Gazetteer files: 2010, 2000, and 1990" (http://www.census.gov/geo/www/gazetteer/gazette.html). United States Census Bureau. 2011-02-12. . Retrieved 2011-04-23.

[5] ghosttowns.com (http://www.ghosttowns.com/states/or/canyoncity.html)

External links

- Canyon City (http://www.gcoregonlive.com/bus_display.php/133) by the Grant County Chamber of Commerce
- Listing for Canyon City (http://bluebook.state.or.us/local/cities/ad/canyoncity.htm) in the *Oregon Blue Book*

U.S._Route_395_in_Oregon

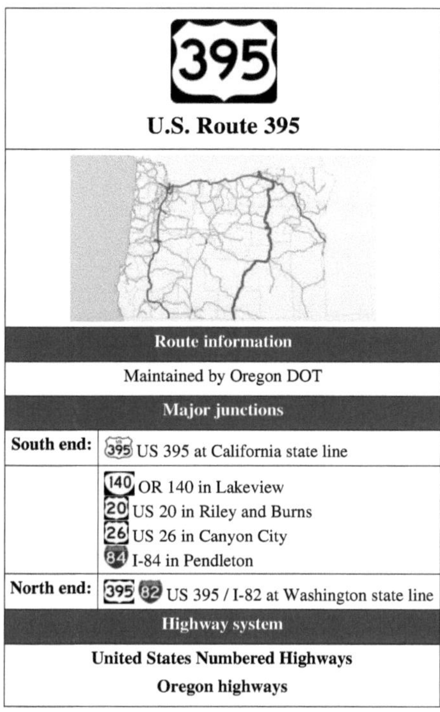

U.S. Route 395

Route information	
Maintained by Oregon DOT	
Major junctions	
South end:	US 395 at California state line
	OR 140 in Lakeview US 20 in Riley and Burns US 26 in Canyon City I-84 in Pendleton
North end:	US 395 / I-82 at Washington state line
Highway system	
United States Numbered Highways Oregon highways	

In the U.S. state of Oregon, U.S. Route 395 traverses the desert and rural areas on the eastern side of the state. The largest cities that US 395 passes through are Pendleton, the county seat of Umatilla County, and Hermiston, also in Umatilla County.

Major intersections

Note: mileposts do not reflect actual mileage due to realignments.

County	Location	Mile[1]	Destinations	Notes
Lake	New Pine Creek	157.73	**California state line**	
	Lakeview	143.03	OR 140 west – Bly, Klamath Falls	South end of OR 140 overlap
		138.34	OR 140 east – Adel, Plush, Winnemucca	North end of OR 140 overlap
	Valley Falls	120.57 90.02	OR 31 – Paisley, Silver Lake, Bend	

County	Location	Mile markers	Destinations	Notes
Harney	Riley	0.11 104.77	US 20 west – Bend, Redmond, Portland	South end of US 20 overlap
	Burns	131.50	OR 78 – Crane, Winnemucca	
		134.29 67.61	US 20 east – Vale, Ontario	North end of US 20 overlap
Grant	Canyon City	0.00 162.29	US 26 east – Prairie City, Vale, Baker City	South end of US 26 overlap
	Mount Vernon	154.03 120.51	US 26 west – Dayville, Prineville, Portland	North end of US 26 overlap
	Long Creek	90.26	OR 402 – Hamilton, Monument, Kimberly	
Umatilla		49.57	OR 244 east – Ukiah, La Grande	
	Nye	23.65	OR 74 west – Heppner, Condon	
	Pendleton	1.57 209.17	I-84 east / OR 37 north, Pendleton City Center	Interchange; south end of I-84 overlap; south end of freeway
		207.47	US 30 east – Eastern Oregon Regional Airport, West Pendleton, Pendleton City Center	South end of US 30 overlap
		202.87	Barnhart Road – Stage Gulch	
		199.53	Yoakum Road – Stage Gulch	
		198.54	Lorenzen Road, McClintock Road	
		193.53	Echo Road – Echo, Lexington	Former Lexington-Echo Highway
	Stanfield	188.98 12.68	I-84 west / US 30 west (Lexington-Echo Highway) – Boardman, Echo	Interchange; north end of I-84 / US 30 overlap; north end of freeway
	Hermiston	4.83	OR 207 – Heppner, Walla Walla	
	Umatilla	0.04 184.87	US 730 east – Walla Walla, McNary Dam	South end of US 730 overlap
		184.40 0.79	I-82 east to I-84 / US 730 west – Pendleton, Portland, Umatilla, Irrigon	Interchange; north end of US 730 overlap; south end of I-82 overlap
		0.00	**Umatilla Bridge over the Columbia River (state line)**	

References

[1] Oregon Department of Transportation, Public Road Inventory (http://www.oregon.gov/ODOT/TD/TDATA/rics/PublicRoadsInventory.shtml) (primarily the Digital Video Log), accessed April 2008

Malheur_National_Forest

<table>
<tr><td colspan="2" align="center">Malheur National Forest</td></tr>
<tr><td colspan="2" align="center">IUCN Category VI (Managed Resource Protected Area)</td></tr>
<tr><td colspan="2" align="center">Strawberry Lake in Malheur National Forest</td></tr>
<tr><td>Location</td><td>Oregon, USA</td></tr>
<tr><td>Nearest city</td><td>Canyon City, Oregon</td></tr>
<tr><td>Coordinates</td><td>44°17′00″N 118°47′04″W[1]</td></tr>
<tr><td>Area</td><td>1700000 acres (6880 km^2)</td></tr>
<tr><td>Visitors</td><td>242,000 (in 2006)[2]</td></tr>
</table>

The **Malheur National Forest** is a National Forest in the U.S. state of Oregon. It contains 1.7 million acres (6900 km²) in the Blue Mountains of eastern Oregon. The forest include high desert grasslands, sage, juniper, pine, fir, and other tree species. Elevations vary from about 4000 feet (1200 m) to the 9038 foot (2754 m) peak of Strawberry Mountain. The Strawberry Mountains extend east to west through the center of the forest. U.S. Route 395 runs south to north through the forest, while U.S. Route 26 runs east-west.

Monument Rock in Malheur NF

The forest is managed for timber extraction, cattle grazing, gold mining and wilderness use by the Forest Service, a division of the US Department of Agriculture. A 1993 Forest Service study estimated that the extent of old growth in the forest was 312000 acres (ha)[3] .

The forest was established by President Theodore Roosevelt on June 13, 1908 and is named after the Malheur River, from the French, meaning literally "misfortune".

In descending order of land area the forest is located in parts of Grant, Harney, Baker, and Malheur counties. There are three ranger districts in the forest, with offices in John Day, Prairie City, and Hines, Oregon.

The Malheur National Forest contains the largest known organism (by area) in the world: an *Armillaria solidipes* that spans 2200 acres (8.9 km²).[4]

Wilderness

There are two wilderness areas in Malheur National Forest.

- Strawberry Mountain Wilderness at 68700 acres (278 km^2)
- Monument Rock Wilderness at 19620 acres (79 km^2), located partially within the Wallowa–Whitman National Forest

See also

- Vinegar Hill-Indian Rock Scenic Area

References

[1] "Malheur National Forest" (http://geonames.usgs.gov/pls/gnispublic/f?p=gnispq:3:::NO::P3_FID:1155108). Geographic Names Information System, U.S. Geological Survey. . Retrieved 2009-04-11.

[2] "Revised Visitation Estimates" (http://www.fs.fed.us/recreation/programs/nvum/revised_vis_est.pdf). National Forest Service. .

[3] Bolsinger, Charles L.; Waddell, Karen L. (1993). *Area of old-growth forests in California, Oregon, and Washington* (http://www.fs.fed.us/pnw/pubs/pnw_rb197.pdf). United States Forest Service, Pacific Northwest Research Station. Resource Bulletin PNW-RB-197.

[4] Beale, Bob. 10 April 2003. Humungous fungus: world's largest organism? (http://www.abc.net.au/science/news/enviro/EnviroRepublish_828525.htm) at Environment & Nature News, ABC Online. Accessed 25 June 2008.

External links

- Malheur National Forest home page (http://www.fs.fed.us/r6/malheur)
- Grant County Chamber of Commerce Information (http://www.grantcounty.cc/zones/usfs/malheur/general/index.php)
- Forest Service home page (http://www.fs.fed.us/)
- Department of Agriculture home page (http://www.usda.gov/wps/portal/usdahome)
- Cedar Grove Botanical Area (http://blog.conifercountry.com/2010/08/23/cedar-grove-botanical-area--aldrich-mountains-oregon.aspx)

Oregon_and_Northwestern_Railroad

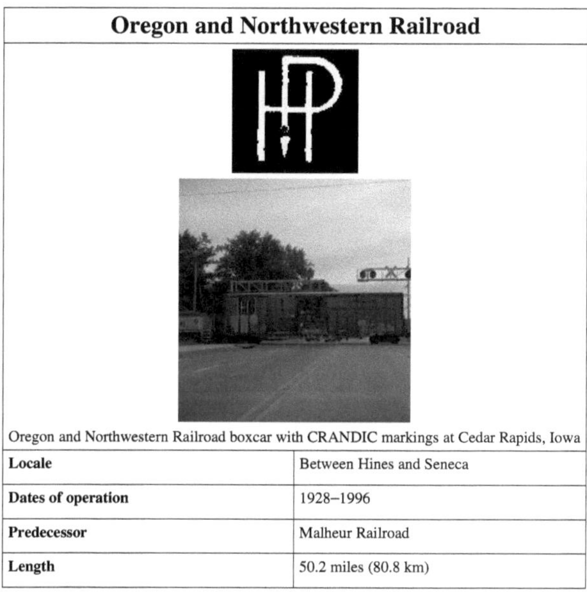

Oregon and Northwestern Railroad

Oregon and Northwestern Railroad boxcar with CRANDIC markings at Cedar Rapids, Iowa

Locale	Between Hines and Seneca
Dates of operation	1928–1996
Predecessor	Malheur Railroad
Length	50.2 miles (80.8 km)

The **Oregon and Northwestern Railroad** (O&NW) evolved from the defunct Malheur Railroad in 1928 and eventually ran 50.2 miles (80.8 km) between Hines, Oregon, and Seneca,[1] [2] along the present-day route of the U.S. 395 in Oregon. The Edward Hines Lumber Company, or Edward Hines Western Pine Company, purchased the Malheur Railroad from Fred Herrick in 1928 and expanded the railroad's network in order to make the company's lumber mill more easily accessible from logging locations. The company planned to log 120000000 board feet (m^3) of lumber each year.[1] [3]

The railroad received its permit from the Interstate Commerce Commission and became a common carrier on 24 June 1929. Charles John Pettibone was the superintendent of the railroad and assistant manager of the lumber company.[3] [4]

In the 1940s, most of the logging was done in order to supply airplanes for the U.S. Army. During this time, the FBI had been deporting Japanese Americans following Japan's attack on Pearl Harbor, but the lumber company needed new workers. Most of the company's original workers had gone to war. The company began to recruit Japanese Americans for the railroad. Each person was obliged "to swear an oath of loyalty to the United States" before beginning to work.[5]

The lumber company owned and operated the railroad, which had 19 trestles,[5] for many decades, but by December 1981, demand for lumber had declined. The company was producing at only one quarter of capacity and employed only 12 workers for the railroad. At the time, it had 229 total employees, which was nearly four times fewer than its peak number, 900. Many workers had been laid off in 1980.[6] The railroad was completely abandoned in 1990 because of damage from the flooding of Malheur Lake and because it was no longer profitable for the logging industry.[7] Four years later, in 1994, the railroad's 475-foot (145 m) tunnel, which had not been used since 1984, was closed to the public because its ceiling was beginning to collapse.[8] Although the tracks were not well-built either,[5] the railroad has been well-preserved.[9]

References

[1] Repp, T. O. (1989). *Main Streets of the Northwest* (http://books.google.com/?id=_tIeAQAAIAAJ&q=Main+Streets+of+the+ Northwest:+Oregon.+Idaho.+Montana&dq=Main+Streets+of+the+Northwest:+Oregon.+Idaho.+Montana). Trans-Anglo Books. p. 78. ISBN 9780870460852. .

[2] Lewis, Edward A. (1996). *American Shortline Railway Guide* (http://books.google.com/?id=3i6K_Nf9e2EC&pg=PA360&dq=Oregon+ and+Northwestern+Railroad#v=onepage&q=Oregon and Northwestern Railroad&f=false) (Fifth ed.). Waukesha, Wis.: Kalmbach Publishing. p. 360. ISBN 0890242909. .

[3] Oregon News Bureau (25 June 1929). "Permit given railway" (http://libweb.uoregon.edu/dc/newspaper/article. php?newspaper=oregonian&id=441505). *The Oregonian* (Portland, Ore.): p. 3. . Retrieved 18 June 2011.

[4] Society of American Military Engineers (1930). *Directory of Members, Constitution and By-laws of the Society of American Military Engineers* (http://books.google.com/?id=peJAAQAAIAAJ&q="Oregon+and+Northwestern+Railroad"&dq="Oregon+and+ Northwestern+Railroad"). **22**. Society of American Military Engineers. . Retrieved 17 June 2011.

[5] "The story of Trout Creek Camp" (http://nl.newsbank.com/nl-search/we/Archives?p_product=MTOR&p_multi=MTEB& p_theme=mtor&p_action=search&p_maxdocs=200&p_topdoc=1&p_text_direct-0=12B168AAE624DCB0& p_field_direct-0=document_id&p_perpage=10&p_sort=YMD_date:D&s_trackval=GooglePM). *The Blue Mountain Eagle* (John Day, Ore.). 1 October 2009. . Retrieved 17 June 2011.

[6] "Partial reopening stirs optimism at lumber mill" (http://news.google.com/newspapers?id=Q5BTAAAAIBAJ&sjid=t4YDAAAAIBAJ& pg=4460,64464&dq=oregon-and-northwestern-railroad&hl=en). *The Bulletin* (Bend, Oregon): p. B3. 29 December 1981. . Retrieved 17 June 2011.

[7] Schwieterman, Joseph P. (2004). *When the Railroad Leaves Town* (http://books.google.com/?id=kzk_m4P9Y6kC&printsec=frontcover& dq=When+the+railroad+leaves+town:+American+communities+in+the+age+of+rail+line+abandonment#v=onepage&q&f=false). Kirksville, Mo.: Truman State University Press. p. 228. ISBN 1931112142. .

[8] "Historic railroad tunnel ruled unsafe, shut to public". *The Oregonian*: p. C2. 1 September 1994.

[9] Edwards, Brian. "Burns to Seneca: The Oregon and Northwestern Railroad" (http://www.abandonedrails.com/ Oregon_and_Northwestern_Railroad). Abandoned Rails. . Retrieved 17 June 2011.

External links

- Photo of O&NW car (http://rickrrpicture.rrpicturearchives.net/showPicture.aspx?id=2321276) by Richard Gibson
- Photos related to the railroad (http://www.flickr.com/search/?q=Oregon+Northwestern+Railroad&ct=6& mt=all&adv=1) from Flickr

Pinus ponderosa

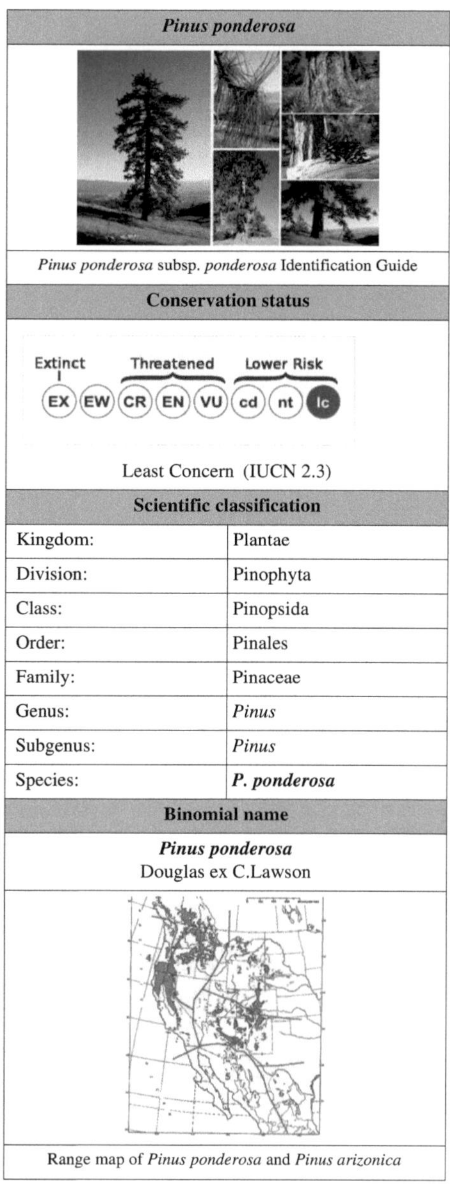

Pinus ponderosa	
Pinus ponderosa subsp. *ponderosa* Identification Guide	
Conservation status	
Least Concern (IUCN 2.3)	
Scientific classification	
Kingdom:	Plantae
Division:	Pinophyta
Class:	Pinopsida
Order:	Pinales
Family:	Pinaceae
Genus:	*Pinus*
Subgenus:	*Pinus*
Species:	***P. ponderosa***
Binomial name	
Pinus ponderosa Douglas ex C.Lawson	
Range map of *Pinus ponderosa* and *Pinus arizonica*	

Pinus ponderosa, commonly known as the **Ponderosa Pine**, **Bull Pine**, **Blackjack Pine**,[1] or **Western Yellow Pine**, is a widespread and variable pine native to western North America. It was first described by David Douglas in 1826, from eastern Washington near present-day Spokane. The ponderosa pine is the official state tree of the State of

Montana.

Distribution

Pinus ponderosa is a dominant tree in the Kuchler plant association, the Ponderosa shrub forest. Like most western pines, the ponderosa is associated with mountainous topography. It is found on the Black Hills and on foothills and mid-height peaks of the northern, central and southern Rocky Mountains as well as the Cascades and Sierra Nevada, and the Maritime Coast Range Ponderosa Pine forests.

Description

Bark, Yosemite National Park, CA

The Ponderosa Pine has a very distinct bark. Unlike most conifers, it has an orange bark, with black lining the crevasses, where the bark "splits". This is most noticeable amongst older Ponderosa Pines. The tree can often be identified by its characteristic long needles that grow in tufts of two or three, depending on subspecies. Its needles are also the only known food of the caterpillars of the gelechiid moth *Chionodes retiniella.*

Sources differ on the scent of the Ponderosa Pine. Some state that the Ponderosa Pine has no distinctive scent,[2] while others state that the bark smells like vanilla if sampled from a furrow of the bark.[3] Sources agree that the Jeffrey Pine is more strongly scented than the Ponderosa Pine.[2] [4]

Size

The National Register of Big Trees lists a Ponderosa Pine that is 235 ft (72 m) tall and 324 in (820 cm) in circumference.[5] In January 2011, a Pacific Ponderosa Pine in Siskiyou National Forest in Oregon was measured with a laser to be 268.35 ft (81.79 m) high. The measurement was performed by Michael Taylor, and Mario Vaden, a professional arborist from Oregon. The tree was climbed on October 13, 2011 by Ascending The Giants (a tree climbing company in Portland, Oregon) and directly measured with tape-line at 268.29 ft (81.77 m) high. [6] [7] This Ponderosa Pine is now the tallest known pine. The previous tallest known pine was a Sugar Pine.

Taxonomy

Modern forestry research identifies four different taxa of Ponderosa Pine, with differing botanical characters and adapted to different climatic conditions. These have been termed "geographic races" in forestry literature, while some botanists historically treated them as distinct species. In modern botanical usage, they best match the rank of subspecies, but not all of the relevant botanical combinations have been formally published.

Subspecies

1. *Pinus ponderosa* subsp. *ponderosa* Douglas ex C. Lawson - **North Plateau Ponderosa Pine**.

   - Range & climate: southeast British Columbia, Washington and Oregon east of the Cascade Range, Arizona, northwestern Nevada, Idaho and western Montana. Cool, relatively moist summers; very cold, snowy winters (except in the very hot and very dry summers of central Oregon, most notably near Bend, which also has very cold and generally dry winters).

2. *Pinus ponderosa* subsp. *scopulorum* (Engelm.) E. Murray - **Rocky Mountains Ponderosa Pine**.

   - Range & climate: eastern Montana, North Dakota, South Dakota, Wyoming, Nebraska, northern and central Colorado and Utah, and eastern Nevada. Warm, relatively dry summers; very cold, fairly dry winters.

Subspecies *scopulorum*, Custer State Park, SD

3. *Pinus brachyptera* Engelm. - **Southwestern Ponderosa Pine**

   - Range & climate: southern Colorado, southern Utah, northern and central New Mexico and Arizona, and westernmost Texas. The Gila Wilderness contains one of the world's largest and healthiest forests.[8] Hot, relatively moist summers; mild winters.

4. *Pinus benthamiana* Hartw. - **Pacific Ponderosa Pine**

   - Range & climate: Washington and Oregon west of the Cascade Range, California, and just into westernmost Nevada. Hot, dry summers; mild wet winters.

The distributions of the subspecies, and that of the closely related Arizona Pine (*Pinus arizonica*) are shown on the map. The numbers on the map correspond to the taxon numbers above and in the table below. The base map of the species range is from Critchfield & Little, *Geographic Distribution of the Pines of the World*, USDA Forest Service Miscellaneous Publication 991 (1966).

Before the distinctions between the North Plateau race and the Pacific race were fully documented, most botanists assumed that Ponderosa Pines in both areas were the same. So when two botanists from California found a distinct tree in western Nevada in 1948 with some marked differences from the Ponderosa Pine they were familiar with in California, they described it as a new species, Washoe Pine, *Pinus washoensis*. However, subsequent research has shown that this is merely a southern outlier of the typical North Plateau race of Ponderosa Pine.

Distinguishing subspecies

Taxon	1 North Plateau	2 Rocky Mts	3 Southwest	4 Pacific	5 Arizona	6 Storm's
Character	(*ponderosa*)	(*scopulorum*)	(*brachyptera*)	(*benthamiana*)	(*arizonica*)	(*stormiae*)
Needles per fascicle	**3**	2-**3**	2-**3**	**3**	**4**-5	3-**5**
Needle length	10–22 cm	8–17 cm	12–21 cm	15–30 cm	12–22 cm	20–30 cm
Needle thickness	1.7-2.2 mm	1.5-1.7 mm	1.6-1.9 mm	1.3-1.7 mm	1.0-1.1 mm	1.0-1.2 mm
Cone length	5–11 cm	5–9 cm	5–10 cm	7–16 cm	5–9 cm	6–11 cm
Cone scale width	14–19 mm	16–20 mm	14–19 mm	18–23 mm	15–18 mm	12–17 mm
Immature cone colour	purple	green	green	green	green	green
Mature cone surface	matte	matte	glossy	glossy	glossy	matte
Seedwing to seed length ratio	1.9-2.5	2.1-3.4	3.0-3.5	3.0-4.7	2.8-3.2	3.0-3.5
Max tree height	50 m	40 m	50 m	81 m	35 m	20 m
USDA hardiness zone	4	4	6	7	7	8

Notes:

Taxon numbers refer to the map

Needles per fascicle - the most frequent number is in **bold**

Seedwing : seed length ratio - high numbers indicate a small seed with a long wing; low numbers a large seed with a short seedwing

Threats

Blue stain fungus, *Grosmannia clavigera*, attacks this species from the mouth of the Mountain Pine Beetle.

See also

- Ponderosa shrub forest
- Maritime Coast Range Ponderosa Pine forests

Notes

[1] Moore, Gerry; Kershner, Bruce; Craig Tufts; Daniel Mathews; Gil Nelson; Spellenberg, Richard; Thieret, John W.; Terry Purinton; Block, Andrew (2008). *National Wildlife Federation Field Guide to Trees of North America*. New York: Sterling. p. 89. ISBN 1-4027-3875-7.

[2] Schoenherr, Allan A (1995). *A Natural History of California*. University of California Press. p. 111.

[3] Kricher, John C (1998). *A field guide to Rocky Mountain and southwest forests*. Houghton Mifflin. p. 194.

[4] Kricher, John C. (1998). *A field guide to California and Pacific Northwest forests*. Houghton Mifflin. p. 107.

[5] "Pacific ponderosa Pine" (http://www.americanforests.org/resources/bigtrees/register.php?details=3961). *National Register of Big Trees*. American Forests. .

[6] Gymnosperm Database - Pinus Ponderosa benthamiana (http://www.conifers.org/pi/Pinus_ponderosa_benthamiana.php)

[7] Fattig, Paul (2011-01-23). "Tallest of the tall" (http://www.mailtribune.com/apps/pbcs.dll/article?AID=/20110123/NEWS/101230353/). *Mail Tribune* (Medford, Oregon). . Retrieved 2011-01-27.

[8] Arizona Mountains forests (http://www.worldwildlife.org/wildworld/profiles/terrestrial/na/na0503_full.html) at World Wildlife Fund.

References

- Conifer Specialist Group (1998). *Pinus ponderosa*. 2006. *IUCN Red List of Threatened Species*. IUCN 2006. www.iucnredlist.org (http://www.iucnredlist.org). Retrieved on 12 May 2006.
- Baumgartner, D. M. & Lotan, J. E. (eds.) (1988). *Ponderosa Pine the species and its management*. Symposium proceedings. Cooperative Extension, Washington State University.
- Conkle, M. T. & Critchfield, W. B. (1988). Genetic Variation and Hybridization of Ponderosa Pine. Pp. 27–44 in Baumgartner, D. M. & Lotan, J. E. (eds.).
- Critchfield, W. B. (1984). Crossability and relationships of Washoe Pine. *Madroño* 31: 144-170.
- Farjon, A. (2nd ed., 2005). *Pines*. Brill, Leiden & Boston. ISBN 90-04-13916-8.
- Haller, J. R. (1961). Some recent observations on Ponderosa, Jeffrey and Washoe Pines in Northeastern California. *Madroño* 16: 126-132.
- Haller, J. R. (1965). Pinus washoensis in Oregon: taxonomic and evolutionary implications. *Amer. J. Bot.* 52: 646.
- Haller, J. R. (1965). The role of 2-needle fascicles in the adaptation and evolution of Ponderosa Pine. *Brittonia* 17: 354-382.
- Lauria, F. (1991). Taxonomy, systematics, and phylogeny of *Pinus* subsection *Ponderosae* Loudon (Pinaceae). Alternative concepts. *Linzer Biol. Beitr.* 23 (1): 129-202.
- Lauria, F. (1996). The identity of *Pinus ponderosae* Douglas ex C.Lawson (Pinaceae). *Linzer Biol. Beitr.* 28 (2): 99-1052.
- Lauria, F. (1996). Typification of *Pinus benthamiana* Hartw. (Pinaceae), a taxon deserving renewed botanical examination. *Ann. Naturhist. Mus. Wien* 98 (B Suppl.): 427-446.
- Smith, R. H. (1977). Monoterpenes of Ponderosa Pine xylem resin. *USDA Tech. Bull.* 1532.
- Smith, R. H. (1981). Variation in Immature Cone Color of Ponderosa Pine (Pinaceae) inNorthern California and Southern Oregon. *Madroño* 28: 272-274.
- Van Haverbeke, D. F. (1986). Genetic Variation in Ponderosa Pine: A 15-Year Test of Provenances in the Great Plains. *USDA Forest Service Research Paper* RM-265.
- Wagener, W. W. (1960). A comment on cold susceptibility of Ponderosa and Jeffrey Pines. *Madroño* 15: 217-219.

External links

- USDA Plants Profile: *Pinus ponderosa* (http://plants.usda.gov/java/profile?symbol=PIPO)
- Gymnosperm Database: *Pinus ponderosa* (http://www.conifers.org/pi/pin/ponderosa.htm)
- Jepson Manual treatment - *Pinus ponderosa* (http://ucjeps.berkeley.edu/cgi-bin/get_JM_treatment. pl?195,210,216)
- *Pinus ponderosa* - Photo Gallery (http://calphotos.berkeley.edu/cgi/img_query?query_src=photos_index& where-taxon=Pinus+contorta)

koi:Pinus ponderosa

Hines,_Oregon

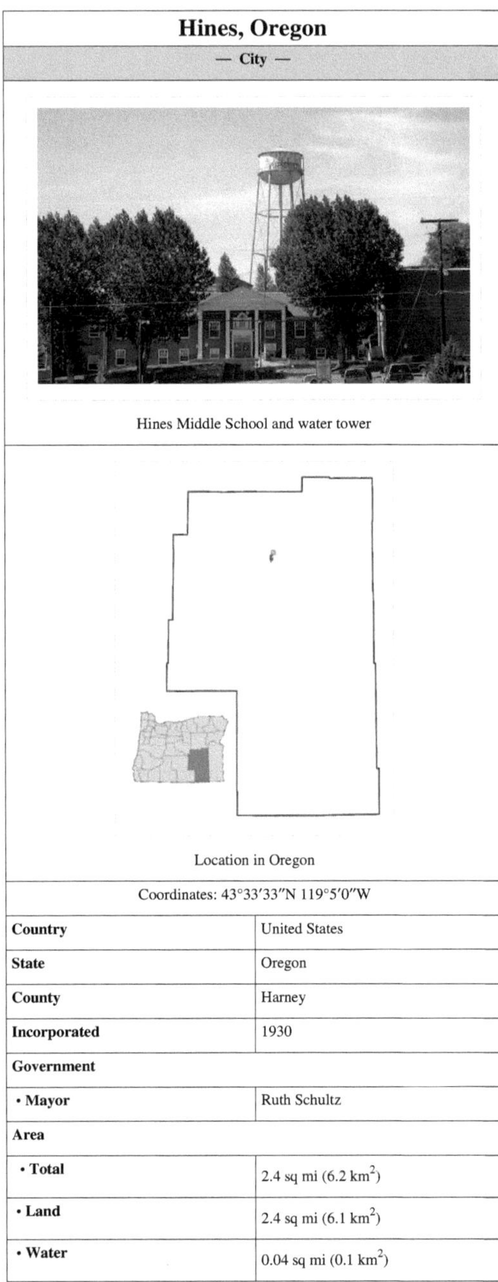

<table>
<tr><td colspan="2" align="center">Hines, Oregon</td></tr>
<tr><td colspan="2" align="center">— City —</td></tr>
</table>

Hines Middle School and water tower

Location in Oregon

Coordinates: 43°33′33″N 119°5′0″W

Country	United States
State	Oregon
County	Harney
Incorporated	1930
Government	
• **Mayor**	Ruth Schultz
Area	
• **Total**	2.4 sq mi (6.2 km^2)
• **Land**	2.4 sq mi (6.1 km^2)
• **Water**	0.04 sq mi (0.1 km^2)

Elevation	4155 ft (1266.4 m)
Population (2010)	
• **Total**	1563
• **Density**	684.8/sq mi (264.4/km^2)
Time zone	Pacific (UTC-8)
• **Summer (DST)**	Pacific (UTC-7)
ZIP code	97738
Area code(s)	541
FIPS code	41-34250[1]
GNIS feature ID	1121843[2]
Website	www.hinesoregon.com [3]

Hines is a city in Harney County, Oregon, United States. The population was 1,563 at the 2010 census.

History

A community named Herrick was formed just southwest of Burns when railroad promoter and sawmill operator Fred Herrick founded a lumber company there.[4] Edward Hines bought the railroad and lumber company from Herrick in 1928, and a post office named Hines was established in 1931 to serve the Edward Hines Lumber Company mill and surrounding community.[4] The mill has since changed hands at least two more times.[4]

Geography

According to the United States Census Bureau, the city has a total area of 2.4 square miles (6.2 km^2), of which 2.4 square miles (6.2 km^2) is land and 0.04 square miles (0.10 km^2) (0.84%) is water.

Demographics

As of the census[1] of 2000, there were 1,623 people, 641 households, and 473 families residing in the city. The population density was 684.8 people per square mile (264.4/km²). There were 689 housing units at an average density of 290.7 per square mile (112.2/km²). The racial makeup of the city was 94.39% White, 0.12% African American, 2.71% Native American, 0.55% Asian, 0.06% Pacific Islander, 0.12% from other races, and 2.03% from two or more races. Hispanic or Latino of any race were 1.66% of the population.

There were 641 households out of which 31.8% had children under the age of 18 living with them, 63.5% were married couples living together, 6.4% had a female householder with no husband present, and 26.2% were non-families. 21.8% of all households were made up of individuals and 8.7% had someone living alone who was 65 years of age or older. The average household size was 2.50 and the average family size was 2.89.

In the city the population was dispersal was 26.6% under the age of 18, 6.3% from 18 to 24, 27.0% from 25 to 44, 26.1% from 45 to 64, and 14.0% who were 65 years of age or older. The median age was 40 years. For every 100 females there were 106.5 males. For every 100 females age 18 and over, there were 98.7 males. The median income for a household in the city was $40,917, and the median income for a family was $43,452. Males had a median income of $32,772 versus $22,458 for females. The per capita income for the city was $15,783. About 6.6% of families and 9.9% of the population were below the poverty line, including 10.1% of those under age 18 and 17.3% of those age 65 or over.

References

[1] "American FactFinder" (http://factfinder.census.gov). United States Census Bureau. . Retrieved 2008-01-31.

[2] "US Board on Geographic Names" (http://geonames.usgs.gov). United States Geological Survey. 2007-10-25. . Retrieved 2008-01-31.

[3] http://www.hinesoregon.com

[4] McArthur, Lewis A.; McArthur, Lewis L. (2003) [First published 1928]. *Oregon Geographic Names* (7th ed.). Portland, Oregon: Oregon Historical Society Press. p. 469. ISBN 9780875952772. OCLC 53075956.

External links

- Hines entry (http://bluebook.state.or.us/local/cities/ek/hines.htm) in the *Oregon Blue Book*
- City of Hines Profile (http://www.harneycounty.org/cityofhines.html) from Harney County Economic Development

Silvies_River

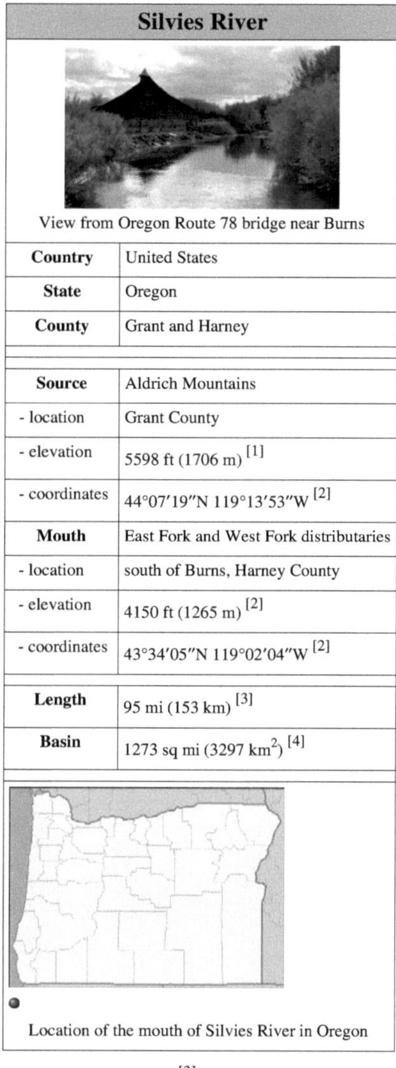

Silvies River

View from Oregon Route 78 bridge near Burns

Country	United States
State	Oregon
County	Grant and Harney
Source	Aldrich Mountains
- location	Grant County
- elevation	5598 ft (1706 m) [1]
- coordinates	44°07′19″N 119°13′53″W [2]
Mouth	East Fork and West Fork distributaries
- location	south of Burns, Harney County
- elevation	4150 ft (1265 m) [2]
- coordinates	43°34′05″N 119°02′04″W [2]
Length	95 mi (153 km) [3]
Basin	1273 sq mi (3297 km^2) [4]

Location of the mouth of Silvies River in Oregon

The **Silvies River** flows for about 95 miles (153 km)[3] through Grant and Harney counties in the U.S. state of Oregon. The river drains 1273 square miles (3300 km^2) of the northern Harney Basin.[4] [5]

The headwaters are on the southern flank of the Aldrich Mountains, about 10 miles (16 km) south of Mount Vernon in Grant County. Named tributaries include Bear Creek and Emigrant Creek. The Silvies runs generally southward and passes near Seneca and Burns. Southeast of Burns, in Harney County, the river splits into two distributaries, the East Fork Silvies River and the West Fork Silvies River. Both terminate at Malheur Lake about 25 miles (40 km) southeast of Burns.[6]

See also

- List of rivers of Oregon
- List of longest streams of Oregon
- Malheur National Wildlife Refuge

References

[1] Source elevation derived from Google Earth search using GNIS source coordinates.

[2] "Silvies River" (http://geonames.usgs.gov/pls/gnispublic/f?p=gnispq:3:::NO::P3_FID:1149516). *Geographic Names Information System (GNIS)*. United States Geological Survey (USGS). November 28, 1980. . Retrieved October 24, 2010.

[3] Sheehan, p. 290

[4] "Silvies – 17120002 8-Digit Hydrologic Unit Profile" (ftp://ftp-fc.sc.egov.usda.gov/OR/HUC/basins/highdesert/17120002_2-23-06.pdf) (PDF). Natural Resources Conservation Service (NRCS). 2006. p. 1. . Retrieved October 24, 2010.

[5] "Silvies River and Tributaries" (http://www.nww.usace.army.mil/dpn/images/silvies-map1.htm) (map). U.S. Army Corps of Engineers. 2005. . Retrieved October 24, 2010.

[6] DeLorme (2008). *Oregon Atlas & Gazetteer* (Map). Section 77–78; 81–82. ISBN 978-0-89933-347-2.

Works cited

- Sheehan, Madelynne Diness (2005). *Fishing in Oregon: The Complete Oregon Fishing Guide*, 10th edition. Scappoose, Oregon: Flying Pencil Publications. ISBN 978-0-916473-15-0.

External links

- U.S. Army Corps of Engineers photo gallery of Silvies River (http://www.nww.usace.army.mil/dpn/images/silvies-photos.htm)

Great_Basin

Great Basin	
region	
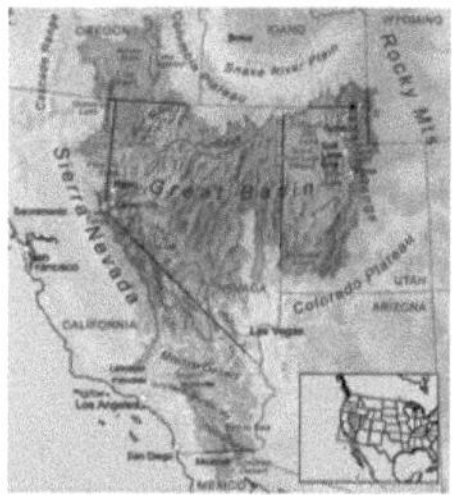 The Great Basin is the multi-state endorheic area surrounded by the Pacific Watershed of North America (Columbia Watershed on the north and the Colorado Watershed on the east and south).	
Countries	United States, Mexico
Location	GNIS-designated point in Sheep Ranch Canyon near Winnemucca, Nevada [1]
- coordinates	40°40′N 117°40′W [1]
Highest point	Mount Whitney summit
- location	Sierra Nevada Mountains
- elevation	14505 ft (4421.1 m)
- coordinates	36°34′42.89″N 118°17′31.18″W
Lowest point	Badwater Basin
- location	Death Valley National Park
- elevation	-282 ft (-86 m)
- coordinates	36°14′23″N 116°50′5″W
Area	184427 sq mi (477664 km^2) [2]
- persistent water (average area)	4079 sq mi (10565 km^2)
Population	3163941
Orogeny	Basin and Range Province fault history
GNIS code	2087988 [1]

Further information: Geography of North America

The **Great Basin** is the largest area of contiguous endorheic watersheds in North America and is noted for its arid conditions and Basin and Range topography that varies from the North American low point at Badwater Basin to the highest point of the contiguous United States, less than 100 miles (160 km) away at the summit of Mount Whitney. The region spans several physiographic divisions, biomes/ecoregions, and deserts, and is the ancestral homeland of the Great Basin tribes.

Geography

The Great Basin includes valleys, basins, lakes, and mountain ranges[3] of Basin and Range topography. The Great Basin almost entirely contains the smaller Great Basin physiographic section, which extends about 10000 sq mi (26000 km^2) into the Colorado River watershed (including the Las Vegas metropolitan area and northwest corner of Arizona). Geographic features near the Great Basin include the Continental Divide of the Americas, the Great Divide Basin, and the Gulf of California.

The Great Basin's two most populous metropolitan areas are Reno to the west and Salt Lake City on the east side. The area between these two cities is sparsely populated, but includes the smaller cities of Delta, Elko, Ely, Tonopah, Wendover, West Wendover, and Winnemucca. The southern area of the basin has the communities of Palmdale, Victorville, and Palm Springs. Major roadways traversing the Great Basin include Interstate 80, Interstate 15, U.S. Route 6, U.S. Route 50, U.S. Route 93, U.S. Route 95 and U.S. Route 395, with the Nevada section of U.S. Route 50 nicknamed "The Loneliest Road in America".[4] Railroad transportation routes also pass through Reno and Salt Lake City.

The Great Basin (magenta outline) differs from the Great Basin Province, the Great Basin tribes area, the Basin and Range Province, and the Basin and Range area.

Hydrology

Wah Wah Valley, Utah, thunderstorm

The Great Basin Divide separates the Great Basin from the watersheds draining to the Pacific Ocean. The southernmost portion of the Great Basin is the watershed area of the Laguna Salada. The Great Basin's longest and largest river is the Bear River of 350 mi (560 km),[5] and the largest single watershed is the Humboldt River drainage of roughly 17000 sq mi (44000 km^2). Most Great Basin precipitation is snow, and the precipitation that neither evaporates nor is taken for human use will sink into groundwater aquifers, while evaporation of collected water occurs from geographic sinks.[6] Lake Tahoe, North America's largest alpine lake,[7] is part of the Great Basin's central Lahontan subregion. Great Basin named deserts include the Black Rock Desert, the Great Salt Lake Desert, the Sevier Desert, the Smoke Creek Desert, Nevada salt deserts in the Great Basin Province, the Mojave Desert, and part of the Sonoran Desert.

Great Basin snowstorm in the Snake Valley of
Utah and Nevada

The Tule Valley watershed and the House Range
(Notch Peak) are part of the Great Basin's Great
Salt Lake hydrologic unit

Ecology

Although mostly within the North American Desert ecoregion, portions of the Great Basin extend into the Forested Mountain and Mediterranean California ecoregions. The semi-arid areas of the Forested Mountain ecoregion include the White Mountains and Inyo Mountains. Extending north of the 42nd parallel north, the Northern Basin and Range (ecoregion) has its southern border at the highest shoreline of the Pleistocene Lake Bonneville. The south edge of the Central Basin and Range ecoregion is in Nevada, north of the south edge of the Great Basin section.

Fauna

Great Basin wildlife includes Pronghorn, Mule Deer, Mountain Lion, and Lagomorphs such as Black-tailed Jackrabbit and Desert Cottontail and the coyotes that prey on them. Packrats, Kangaroo rats and other small rodents are also common, but these are predominantly nocturnal. Elk and Bighorn Sheep are present but uncommon. Small lizards such as the Great Basin fence lizard, Longnose Leopard Lizard and Horned Lizard are common, especially in lower elevations. Rattlesnakes and Gopher snakes are also present. The Inyo Mountains Salamander is endangered. Shorebirds such as Phalaropes and Curlews can be found in wet areas. American White Pelicans are common at Pyramid Lake. Golden Eagles are perhaps more common in the Great Basin than anywhere else in the US. Mourning Dove, Western Meadowlark, Black-billed Magpie, and Common Raven are other common bird species.

Two endangered species of fish are found in Pyramid Lake: the Cui-ui sucker fish and the Lahontan cutthroat trout.[8]

Large invertebrates include tarantulas (Aphonopelma genus) and Mormon crickets. Exotic species, including Chukar, Grey Partridge, and Himalayan Snowcock, have been successfully introduced to the Great Basin, although the latter has only thrived in the Ruby Mountains. Cheatgrass, an invasive species which was unintentionally introduced, forms a critical portion of their diets. Feral horses (Mustangs) and wild burros are other highly reproductive, and ecosystem controversial, alien species. Most of the Great Basin is open range and domestic cattle and sheep are widespread.

Flora

Utah Juniper/Single-leaf Pinyon (southern regions) and Mountain Mahogany (northern regions) form open Pinyon-juniper woodland on the slopes of most ranges. Stands of Limber Pine and Great Basin Bristlecone Pine (Pinus longaeva) can be found in some of the higher ranges (the Methuselah tree is nearly 5000 years old). In riparian areas with dependable water cottonwoods (Populus fremontii) and Quaking Aspen (Populus tremuloides) groves exist. Grasslands contain the native Great Basin wildrye.

History

Sediment build-up over thousands of years between undersea ranges created relatively flat lacustrine plains in underwater portions of the prehistoric area that would be named the Great Basin after they drained.[9] For example, after forming about 32,000 years ago, Lake Bonneville overflowed about 14,500 years ago in the

The Black Rock Desert Volcanic Field of Utah (pictured) differs from the Black Rock Desert of Nevada, and both are in the Great Basin.

Bonneville Flood through Red Rock Pass and lowered to the "Provo Lake"[10] level (the Great Salt Lake, Utah Lake, Sevier Lake, Rush Lake, and Little Salt Lake remain).[11] Lake Lahontan, Lake Manly, and Lake Mojave were similar Pleistocene lakes.

Paleo-Indian habitation by the Great Basin tribes began as early as 10,000 B.C. (the Numic-speaking Shoshonean peoples arrived as late as 1000 A.D.).[12] Archaeological evidence of habitation sites along the shore of Lake Lahontan date from the end of the ice age when its shoreline was approximately 500 feet (150 m) higher along the sides of the surrounding mountains. The Great Basin was inhabited for at least several thousand years by Uto-Aztecan language group-speaking Native American Great Basin tribes, including the Shoshone, Ute, Mono, and Northern Paiute.

Exploration of the Great Basin occurred during the 18th century Spanish colonization of the Americas. The first American to cross the Great Basin from the Sierra Nevada was Jedediah Strong Smith in 1827. [13] Peter Skene Ogden of the British Hudson's Bay Company explored the Great Salt Lake and Humboldt River regions in the late 1820s, following the eastern side of the Sierra Nevada to the Gulf of California.[14] Benjamin Bonneville explored the northeast portion during an 1832 expedition. The United States had acquired control of the area north of the 42nd parallel via the 1819 Adams–Onís Treaty with Spain and 1846 Oregon Treaty with Britain. The US gained control of most of the rest of the Great Basin via the 1848 Mexican Cession. The first non-indigenous settlements were connected with the eastern regions of the 1848 California Gold Rush, with its immigrants crossing the Great Basin on the California Trail along Nevada's Humboldt River to Carson Pass in the Sierras. The first American religious settlement effort was the Mormon provisional State of Deseret in 1849 in present day Utah and northern Nevada. The Oregon Territory was established in 1848 and the Utah Territory in 1850.

In 1869 the First Transcontinental Railroad was completed at Promontory Summit in the Great Basin. Around 1902, the San Pedro, Los Angeles and Salt Lake Railroad was constructed in the lower basin and Mojave Desert for California-Nevada rail service to Las Vegas, Nevada.

To close a 1951 Indian Claims Commission case, the Western Shoshone Claims Distribution Act of 2004 established the United States payment of $117 million to the Great Basin tribe for the acquisition of 39000 square miles (km^2).[15] The Dixie Valley, Nevada, earthquake (6.6-7.1) in the Great Basin was in 1954. The Mojave and Colorado Deserts Biosphere Reserve (4 conserved areas) was designated in 1984, and in 1986, and 1994, the Great Basin National Park and the Mojave National Preserve were established. In 2009, the American Land Conservancy's Great

Basin Program reserved the Green Gulch mule deer migration corridor as part of *"over 80,000 acres* [conserved] *in Nevada and the Eastern Sierras"*.[16]

References

[1] "Great Basin (2087988)" (http://geonames.usgs.gov/pls/gnispublic/f?p=gnispq:3:::NO::P3_FID:2087988). Geographic Names Information System, U.S. Geological Survey. . Retrieved 2011-10-01.

[2] "Boundary Descriptions and Names of Regions, Subregions, Accounting Units and Cataloging Units" (http://water.usgs.gov/GIS/huc_name.html). U.S. Geological Survey. . Retrieved 2010-10-23.

[3] "Basin and Range Province" (http://geomaps.wr.usgs.gov/parks/province/basinrange.html). *Geologic Provinces of the United States.* United States Geological Survey. 2004. . Retrieved 2009-01-10.

[4] "The Official Hwy 50 Survival Guide — The Loneliest Road in America" (http://travelnevada.com/documents/guides/survival_guide.pdf) (PDF). Nevada Commission on Tourism. . Retrieved 2007-12-15.

[5] "Bear River Watershed Description" (http://www.bearriverinfo.org/description/). Bear River Watershed Information System. . Retrieved 2010-04-28. (an additional ~1% is in the SW corner of WY)

[6] "Great Basin" (http://www.nature.nps.gov/Geology/usgsnps/province/brgrbas.html). *Geologic Provinces of the United States: Basin and Range Province.* nature.nps.gov: National Park Service. . Retrieved 2009-01-10.

[7] "Amazing Lake Tahoe" (http://www.bluelaketahoe.com/page.php?p=amaz&l=1). Lake Tahoe Visitors Authority. . Retrieved 2008-10-26.

[8] Hogan, C.Michael; Papineau, Marc, et al (1987). *Development of a dynamic water quality simulation model for the Truckee River.* Environmental Protection Agency Technology Series. Washington D.C.: Earth Metrics Inc..

[9] Jackson, Richard H.; Stevens, Dale J. (1981). "Physical and Cultural Environment of Utah Lake and Adjacent Areas" (https://ojs.lib.byu.edu/ojs/index.php/gbnmem/article/view/2941/3289). *Great Basin Naturalist Memoirs* (5: Utah Lake Monograph): 5. . Retrieved 2010-04-06.

[10] Gilbert, Grove Karl (1890) (Google Books). *Lake Bonneville* (http://books.google.com/books?id=NY0sAAAAYAAJ&pg=RA1-PA127). p. 127. . Retrieved 2010-04-23.

[11] Morgan, Dale L (1947). *The Great Salt Lake.* Salt Lake City: University of Utah Press. p. 22. ISBN 0-87480-478-7.

[12] "Archaeology, Cultural Transmission, and the Indigenous Native American Indians of the Great Basin Region of North America" (http://www.bauuinstitute.com/Articles/ArchaeologyCultureGreatBasin.html). Bauu Institute. . Retrieved 2010-04-22.

[13] Morgan (1953, 1964), *Jedediah Smith and the Opening of the West*, p. 7

[14] Ogden, Peter Skene (http://www.biographi.ca/009004-119.01-e.php?&id_nbr=4109), Dictionary of Canadian Biography Online

[15] "Action Alert!" (http://www.shundahai.org/claims_action_alert_0501.htm). Shundahai Network. . Retrieved 2010-04-22.

[16] "Great Basin" (http://www.alcnet.org/projects/overview/basin). *Programs.* American Land Conservancy. . Retrieved 2010-01-11.

Article Sources and Contributors

Seneca,_Oregon *Source*: http://en.wikipedia.org/w/index.php?title=Seneca%2C_Oregon *Contributors*: Ajbenj, Bumm13, Droll, EncMstr, Gianfranco, Hike796, Jllm06, Langard, Lesb246, LilHelpa, MJCdetroit, Mithridates, N.M.Sheedy, Nyttend, Oopopop, Owen, Ram-Man, Rich Farmbrough, Tesscass, Valfontis, 4 anonymous edits

Grant_County,_Oregon *Source*: http://en.wikipedia.org/w/index.php?title=Grant_County%2C_Oregon *Contributors*: AXRL, Ajbenj, Backspace, Bumm13, CanisRufus, CommonsDelinker, D6, DBerend, Dino, Dukeofomnium, Eaglecap Backpack, EncMstr, Finetooth, Floydspinky71, Fratrep, Gaius Cornelius, Ground Zero, Hmains, Hyperfast, Ipoellet, John of Reading, Jsayre64, Katieh5584, Langard, Lesb246, Leslie Mateus, Lightmouse, Lincolnite, Llywrch, Monegasque, Motthoangwehuong, N.M.Sheedy, Northwesterner1, Nyttend, Owen, Pearle, Peteforsyth, PurpleSunflower, Ram-Man, Rich Farmbrough, Sasata, Sgt Pinback, Smallbones, TUF-KAT, Tabletop, Template namespace initialisation script, Tim1357, Tom Morris, Valfontis, Wda, Worldenc, Zinc2005, 13 anonymous edits

Oregon *Source*: http://en.wikipedia.org/w/index.php?title=Oregon *Contributors*: (jarbarf), -Fire-, 16@r, 3.217, 5 albert square, 7george7, 7heesub, A2Kafir, A7xjackass13, ACSE, AF501, AHP, Aaron Schulz, Aaron north, Abbalkea2010, Abby, Abce2, Aboutmovies, AceHarding, Adam Keller, AdamjVogt, Addshore, AdjustShift, Aeusoes1, Ag1246, AgentPeppermint, Ahassan05, Ahoerstemeier, Aivazovsky, Aiyda, Ajbenj, Akendall, Al Silonov, Al-bayda, Alansohn, Alberry, Ale jrb, Alexander S., Alexindigo, AlexiusHoratius, Alexmichalskileslie, Alexwcovington, Algocu, Aliceinlampyland, Aliceiscoo, Alison, Almostmidnight, Alphachimp, Altenmann, Amazonien, Amccrea, American Black bear, AmiDaniel, Amsby17, Andrewlp1991, Andy Marchbanks, Andy120290, Angleterre, Antandrus, Antipex, Antodav2007, Apple1091, Aris Katsaris, Aroldis, Art LaPella, Arthuralee, Artsojourner, Arx Fortis, Ashwinr, Astropithicus, Athelwulf, Aude, AustinRyan, Av99, Averybo, Awotter, Axoplasm, BD2412, BMRR, BSveen, Babygene52, Balcer, Ballista, BalooUrsidae, Bancroft595, Barek, Bart133, Bassbonerocks, Bastin, Bbpen, Beano, Beansbaxter, Bearclawbill, Beaver1believer, Becboobear, Beginning, Beland, Belligero, BenBaker, Bender235, Bennelliott, Bertolc, Best O Fortuna, Beveal, Beyondthislife, Bhound89, Big Smooth, Big iron, Bigcheesebebbs, Biruitorul, Bizxman, Bkonrad, Black Orchid, Blaze33541, Bluedogtn, Bobblewik, Bobomejor, Bodegaard, Bomac, Bongwarrior, Bookgrrl, Booyabazooka, Borderer, Brian0918, Brianga, Broctel, Brutaldeluxe, BryceHarrington, Buaidh, Buddha24, Burgundavia, Butts99, Butwhatdoiknow, CALR, CIreland, CJLL Wright, CO, CSL, CWenger, CWii, Cacophony, Caloy24, Calvin 1998, Can't sleep, clown will eat me, Canadian-Bacon, CanadianLinuxUser, CanadianPenguin, CanisRufus, Canseb, Cantea V, Caponer, Capricorn42, Captain panda, Capuchin the Friar, Caranha, CardinalDan, Carlaude, Carlsotr, Caseyo1992, Cburnett, Cchow2, CenozoicEra, Centrx, CeramicReality, Ceranthor, Chadlupkes, Chawman69, ChrisRuvolo, Chrisbrownluvr60, Chrisoule, Chrisrtait, Chrnath, Chuckgeorge60, Chuggnutt, Chzz, Ciescott, Civil Engineer III, ClairSamoht, Clarkbhm, Clipdude, Cluelessness, CommonsDelinker, Conundrum22, Conversion script, Coolcaesar, CopperSquare, Coredesat, Corey.reece, Corriebertus, Coulraphobic123, Courcelles, Craigaswartout, CrazyMan10101, Crazycomputers, Crd721, Cribbswh, Cromwellt, Cwolfsheep, Cybercobra, Cyde, Cylewithak, CylonCAG, CyranoDeWikipedia, Cyrusxneil, D6, DARTH SIDIOUS 2, DCEdwards1966, DIDouglass, DJ1AM, DRMparadigm, DVD R W, Dabomb87, Dale Arnett, Daniel Royer, DanielCD, Danny, Dark jedi requiem, Darklilac, Dave Cohoe, DaveGorman, Daven200520, Davepape, David Justin, David Schaich, Davidzundel, Davy hopkins@yahoo.com, Dawnseeker2000, Dcornwall, Decltype, Decumanus, Dedicated editor, Deeet, Deflective, DeltaQuad, Demi, Den fjättrade ankan, Dendodge, DerHexer, DickClarkMises, Dino, Dirtyaffliction, Discospinster, Dmsar, Doc glasgow, DomiAllStates, Dommkopf, Doug Coldwell, Doulos Christos, Dr. Blofeld, Draco345, Dre503, Dual Freq, Duck4life, Dufekin, Duff, Dwilso, Dysepsion, E0steven, EOBeav, ERcheck, ESkog, Ed g2s, Editor at Large, Edivorce, Edwardsully, Ehurtley, El C, El Cid, Elano, Elwood j blues, Emeraldcityserendipity, Emhoo, EmilyWolff, Emmett5, Emojones, EncMstr, EncycloPetey, Engineer Bob, EngineerScotty, Entirelybs, Eochaid, Epbr123, Epolk, Epsilon60198, Erall, ErgoSum88, Eric-Wester, Eric33sb, EricSerge, Esprqii, Eu.stefan, Evice, Evrik, Explicit, FCSundae, FactionofReds, Faithlessthewonderboy, Falcon8765, Fang Aili, Fastily, Fastrunner61, Fbolanos, Fcb981, Filiwickers, Finder Steve, Finduilas 09, Firien, Fishing, Flaminsky, Flatterworld, Fondue, Foofighter20x, Freak4jc, Friginator, FriscoKnight, Fritz Saalfeld, Fuhghettaboutit, Funnyhat, Fupa12, GOTU62, Gaius Cornelius, Gdo01, Geologyguy, GeorgeMoney, Gettingtoit, Gfoley4, Gimboid13, Glasscity09, Glendoremus, Gogo Dodo, Good Olfactory, GoodDamon, Google1230, GraemeMcRae, Grafen, Graham87, Graymornings, Greekafella, GregU, Gtwkndhpqu, Guliolopez, Gurch, Gwguffey, Gwynnebaer, Hadal, Haha169, Hajatvrc, Half price, HamburgerRadio, Harmil, Haukurth, Hayaji, HeLmiT, Headwes, Heegoop, HeikoEvermann, Henryodell, Higbvuyb, Highvale, Hiimkylestevens, Hlj, Hmains, Hoefer12, HorizonOmen, Hu12, Hubertfarnsworth, Huntington, Husond, Hut 8.5, IBook of the Revolution, IRP, Ice Cold Beer, Idontgotmilk, IkonicDeath, Imawesomeandilikeit, Imnaha, Imran, Indefatigable, Infoporfin, Inter16, Interiot, Intothewoods29, Iop7789, Ipawn, Ipoellet, Iridescent, IsaacTheSalsaShark, Isoperlagravitans, Isopod, ItsZippy, Ixfd64, J'onn J'onzz, J.delanoy, JFG, JForget, JTM, JYOuyang, Ja 62, JackLumber, Jacob.jose, Jacobuskaminus, Jahiegel, Jakobees, James McNally, Jauhienij, Javert, JayHenry, Jcam, Jcworsley, Jellytime86, Jengod, Jennypalazio, Jeremiestrother, Jesse Milligan, Jfire, Jfricker, Jfurr1981, Jfusion, Jgilhousen, Jgwpdx, Jhendin, Jim1138, JimIrwin, JinJian, Jlhw, Jmchuff, Jms39, Jmtn20, Jnorton7558, JoJoVader12, JoanneB, JoaoRicardo, John K, John254, JohnInDC, JohnLloydScharf, JohnOwens, Johnny Mnemonic, Jojhutton, Jomasecu, Joshbzin, Joyous!, Jrat87, Jsayre64, Ju66l3r, Juliancolton, Julien Deveraux, Jump222, JustBear, Justrob, KGasso, KPH2293, Kaijan, Kaiser matias, Katr67, Kayaker, Kbdank71, KeithH, Ken Gallager, Kevin B12, Kevin Myers, Kevin908, Kiko4564, Kilowattradio, Kimchi.sg, King prime Minister, King256, Kinotama, Kintetsubuffalo, Kkailas, Kkmd, Kkv123, Kman543210, Kollision, Kozuch, Krich, KuBi4K, Kuchiguchi, Kudzu1, Kukini, Kumioko, Kungfuadam, Kuribosshoe, Kuru, Kwamikagami, LGagnon, LJTPerry, Lampsalot, Larrykoen, Lateg, Latitude0116, Lean.joe, LeaveSleaves, LedgendGamer, Leszek Jańczuk, LeviathanMist, Levineps, Lfh, Liface, Lifung, LightSParker, Lightmouse, Lights, Lilac Soul, Lilmiss34, Lincpalex, LineOfBlackCars, Little Mountain 5, Llywrch, Longnskinny, Lookiemookiecookie, Lookinhere, Looxix, Lord British, LouI, Luna Santin, Luong, Luuva, Luxa, M C Y 1008, MJCdetroit, MJDTed, MK8, Macarion, Mack2, Magister Mathematicae, Makaristos, Malo, Manayse1, Manuel Trujillo Berges, Marauder09, MarcoTolo, Marek69, MarkSweep, Martin Kozák, Masterminded, Matt Gies, Matt.T, Matt89520, Mattmcc, Mav, Maximus Fagot, Mdroses, Meadey9, Menah the Great, Merovingian, Messynessie125, Metanoid, Methychroma, Mets501, Mhking, Michael Hardy, Michael Patrick, Michaelmanninghansen, Microtonal, Midgley, Midnight Green, Mightymights, MikeTizzle777, Mikevegas40, Mild Bill Hiccup, Miles530, Mimzy1990, Minesweeper, Mkeranat, Mokgen, Monterey Bay, Mothball, Moverton, Moxy, Mr Accountable, Mr Responsible, Mr.batmanperson, MrBoo, Mrtobacco, Ms2ger, Msaroff, MuZemike, Mwanner, Mxn, N734LQ, NE2, NEICenergy, Naddy, Nagasakisullenhorde, Nagelfar, Nakon, NatureA16, NawlinWiki, NellieBly, Neutrality, Nevin488, NewEnglandYankee, Newport Beach, Niccus, Nicknicknuke, Nightkey, Nihiltres, Niteowlneils, Njoedits, Nkrosse, Nn123645, Nobodyiswatching, Nojam75, NorCalHistory, Northamerica1000, Northwestern guy, Northwesterner1, NorwegianBlue, Nufy8, Nuno Tavares, Nw adelman, Oberonfitch, Oddchump007, Offkilter, Ohnoitsjamie, Oldlaptop321, Olivier, Omega65, Omicronpersei8, OregonD00d, OregonDuck37, OregonRepublican, Oregongrl88, Oregonians, Orphan Wiki, Osterluzei, OverMyHead, Owen, Oxymoron83, Oysterguitarist, Ozette, PC1996, PDXblazers, PM800, PS2pcGAMER, Paleowiki, Pandarr, Pascal.Tesson, Patrick, Paul-L, Paul2.0, Pavel Vozenilek, Pekinpekin, Pennn15, Penubag, Peteforsyth, Peter G Werner, Petercoyl, PeytonWestlake, Pfly, Pgan002, Phantomsteve, Phil Boswell, Philip Trueman, Philipscranage, Piano non troppo, Picaroon, Piledhigheranddeeper, Pilotguy, Pinethicket, Pion, Plasticup, Plowboylifestyle, Poccil, PoliceCaptain, Porce, Portlandium, Postdlf, Pourlaclasse, Presumptive, Puget Sound, Qqqqqq, Quackslikeaduck, Quintote, Qutezuce, Qxz, R'n'B, RG2, RJaguar3, RTC, RVJ, RVRVRVRVRVRVC, Rachelthompson44, Railer 103, Rami.J241, RandalSchwartz, Randlo97, RandomP, RattusMaximus, Rawnblade132, Realm of Shadows, Rebecca, RedWolf, Redtitan, Reinerd, Renato Caniatti, RepublicanJacobite, Retired username, Rettetast, RevTenderBranson, Revas, RexNL, Rhatsa26X, Rich Farmbrough, RichardF, Richardaedwards, Rick Block, Riverstepstonegirl, RjLesch, Rjwilmsi, Rmhermen, Roastytoast, Robfhu, Rocketpastsix, Romanm, Romarin, Ronhjones, Root Beers, Rootbeer, Rorschach, Rossami, Rossdegenstein, Rreagan007, Rror, Rubicon, RubixA+, Ruhrfisch, Rupertslander, Ryan5091, Ryulong, SDC, SHIMONSHA, SJP, SNIyer12, STAREYe, SURIV, Sam Hocevar, Samtheboy, SandyGeorgia, Sango123, Sannse, Saopaulo1, Sarrus, Sasata, Satesclop, SatyrTN, Scarab12, Schonchin, SchuminWeb, Scientizzle, Sciurinæ, ScottMainwaring, Scriberius, Scythian1, Seb az86556, SebastianHelm, Secret (renamed), SecurityAdvisorCode17, Seqsea, Sfmontyo, Shalom Yechiel, Shanem4311, Shanes, Shii, Shoecream, Shoeofdeath, Silverrobes, Simeon, Simon Far, SimonTrew, Singularity, Sionus, Skarebo, SkerHawx, Skookum1, SkyAttacker, Slambo, Sligocki, Slj, Smith03, Snigbrook, Snowolf, Some guy on the Internet, SomeDudeWithAUserName, Someguy1221, Sonett72, Sottolacqua, South Bay, Sowelilitokiemu, Sphincter3, SportingFlyer, Spotteddogsdotorg, Spydrlink, Staffwaterboy, Staplegunther, Starnestommy, Steve Casburn, SteveSims, Steven Walling, Steven91, Stevewonder2, Sturmdrang, Sumsum2010, Sunray, SuperHamster, Supergeekfreak, Svgalbertian, Swollib, Synchronism, Szyslak, T-Bone, TDogg310, TEG24601, TJ Spyke, TMC, TUF-KAT, Ta bu shi da yu, Tartanwizard, TartarSauce, Tedder, Tehbroseff, Template namespace initialisation script, Tesscass, Thatguyflint, The Epopt, The High Fin Sperm Whale, The Obento Musubi, The Thing That Should Not Be, The Transhumanist, TheKMan, TheMightyOrb, Theallpowerfulma, Thespuddy, Thewikipedian, Thingg, Thorwald, Thylacinus cynocephalus, Tide rolls, Tidying Up, Tigen, TimeClock871, Timneu22, Tinosa, Titoxd, Tk47517, Tnock, Toiyabe, Tomruke, TopoChecker, Toxicroak, Toytoy, Tpahl, Tracer9999, Tradnor, Trasel, Travis.Thurston, Tree Biting Conspiracy, Trekphiler, Tresiden, Trimalchio, Triona, Tris189Woot, Truflip99, Tuffluffjimmy, TwinCityIL, Twp, Twsx, Tygar, Tyrol69, Ulmanor, Ulric1313, Umdenken, UncleDouggie, Undelope32, Unyoyega, Ursoc, Uyvsdi, Valfontis, Vanished 6551232, Vary, VegaDark, Velella, Velvetron, Versus22, Vesperclint, Vicenarian, Vicki Rosenzweig, VigilancePrime, Virgil61, Viriditas, Vokkare, Vpuliva, Vrenator, Vsmith, Vtskier2010, Vystrix Nexoth, WJBscribe, WVhybrid, Waggers, Wahkeenah, Walrus heart, Wangi, Wapcaplet, Warpflyght, Wars, Waterjuice, Wayne Slam, Wdflake, Weatherman78, Weregerbil, WestWovles13, WhisperToMe, Wiki alf, Wikiguy008, Wikipelli, Willydick, Winchelsea, Wknight94, Wmahan, Wohuigongfu, Wolfscarr, Wolverineblue, Woohookitty, XRK, Xezbeth, Xxkaenxx, Y5nthon5a, YUL89YYZ, Yamamoto Ichiro, Yamla, YellowMonkey, Yintan, Yortek98123, Youngamerican, ZX81, ZabMilenko, ZacoTaco, Zafiroblue05, Zainubrazvi, Zaui, Zaxname123, Zeimusu, Zeno293, Zntrip, Zoe, Zonath, Zoopazoop, Zzuuzz, A, 1789 anonymous edits

Blue_Mountains_(Oregon) *Source*: http://en.wikipedia.org/w/index.php?title=Blue_Mountains_%28Oregon%29 *Contributors*: Aboutmovies, Bryan Derksen, Bumm13, Decumanus, Docu, EncMstr, Gemini1980, Glacier109, Hmains, Ipoellet, Ixfd64, Jllm06, Linnell, Magioladitis, Muchness, Myasuda, NE2, Night Gyr, Northwesterner1, OregonD00d, Pennn15, RedWolf, SD Martin61, SPUI, Scriberius, Tesscass, Thorwald, Topbanana, Ufwuct, Utopies, Valfontis, Wbfergus, Williamborg, 33 anonymous edits

Canyon_City,_Oregon *Source*: http://en.wikipedia.org/w/index.php?title=Canyon_City%2C_Oregon *Contributors*: AXRL, Aboutmovies, Ajbenj, Bobblewik, Bumm13, Dispenser, EncMstr, Finetooth, Gene Nygaard, Global777, J.delanoy, Jllm06, Langard, Lesb246, MJCdetroit, Mithridates, Motthoangwehuong, N.M.Sheedy, Nyttend, Owen, Ram-Man, Skookum1, Valfontis, 8 anonymous edits

U.S._Route_395_in_Oregon *Source*: http://en.wikipedia.org/w/index.php?title=U.S._Route_395_in_Oregon *Contributors*: AL2TB, Aboutmovies, AvicAWB, Bobjgalindo, Dan ad nauseam, Fbdave, Gill Giller Gillerger, I-10, Imzadi1979, Morriswa, Must eat worms, NE2, Onore Baka Sama, Polaron, Rschen7754, Splat5572, 7 anonymous edits

Malheur_National_Forest *Source*: http://en.wikipedia.org/w/index.php?title=Malheur_National_Forest *Contributors*: AchimP, Atarr, Axcordion, Backspace, Behun, Biglovinb, CarolSpears, Clarkbhm, Clarkcj12, Cmdrjameson, Dino, Droll, Edgar181, EncMstr, Florian Huber, GL, Gioto, Grutness, Hike395, Ipoellet, Jllm06, Jonxwood, Leandrod, Lightmouse, Miguel.v, N.M.Sheedy, NE2, Nopira, Peteforsyth, Pseudomonas, R'n'B, Sasata, SaveTheForests, Utopies, Valfontis, Waacstats, 8 anonymous edits

Oregon_and_Northwestern_Railroad *Source*: http://en.wikipedia.org/w/index.php?title=Oregon_and_Northwestern_Railroad *Contributors*: Darwinek, Feddacheenee, Jsayre64, Lightmouse, Ohconfucius, SJ Morg, Valfontis

Pinus ponderosa *Source*: http://en.wikipedia.org/w/index.php?title=Pinus_ponderosa *Contributors*: Abigail-II, Alansohn, Anna Frodesiak, Antandrus, Arthropod, BD2412, Benjaminb, Bigdoog, BlueCanoe, Breawycker, Buaidh, Burlywood, Casey Duke, Casliber, Clyde frogg, Colchicum, Crazyticktockclock, Curtis Clark, Dabean, DanielZM, Dysmorodrepanis, Edward Waverley, Erik Zachte, First Light, Flyhighplato, Hans-Jürgen Hübner, Helloperson212, Hesperian, Hike395, Impala2009, J.delanoy, Jamidwyer, Jeffrey Mall, Jengod, John Nevard, JohnCub, Jsayre64, Karl Dickman, Kmmontandon, Kukini, L Kensington, Look2See1, Ltvine, MPF, Mav, Menchi, Mmcannis, Moerasbabe, Oxymoron83, Pinethicket, Plumpurple, RCopple, Refinnejann, Rei, Retto12, Richard asr, Rockfang, Rying, Sasata, Scott Burley, Seaphoto, Shanes, Shazen27, T34, TDogg310, Template namespace initialisation script, ThreeWikiteers, Tommy2010, Txomin, Valfontis, Vibrantspirit, West Brom 4ever, William Avery, Woahtoster, Wsiegmund, Yarxia, Zscout370, 108 anonymous edits

Hines,_Oregon *Source*: http://en.wikipedia.org/w/index.php?title=Hines%2C_Oregon *Contributors*: Ajbenj, Bumm13, ESVNSCR, EncMstr, Finetooth, Floydspinky71, Hushpuckena, Jllm06, John Cardinal, Lesb246, MJCdetroit, Mithridates, Nyttend, Owen, Peteforsyth, Ram-Man, Valfontis, 6 anonymous edits

Silvies_River *Source*: http://en.wikipedia.org/w/index.php?title=Silvies_River *Contributors*: Aboutmovies, EncMstr, Finetooth, Hike796, Hmains, Little Mountain 5, Marokwitz, Valfontis, Yworo

Great_Basin *Source*: http://en.wikipedia.org/w/index.php?title=Great_Basin *Contributors*: ALK, Acalamari, AdRock, AdjustShift, Alanc, Alansohn, Amalthea, Amaraiel, Analogdemon, Anderfreude, Andycjp, Anlace, Arakunem, Arcadie, Argyriou, Astropithicus, Auntof6, Austinfidel, Authalic, Avenue, BD2412, Baskaufs, Bassbonerocks, Bayerischermann, Bazonka, Beefyt, Before My Ken, Bob rulz, Bobo192, Bplewe, Brianlucas, Bryan Derksen, Caltas, Carnildo, Carroy, Chris.urs-o, Cmguy777, Coemgenus, Conorobradaigh, Cutler02, Cycotic, D6, DSYoungEsq, Dale101usa, Danny, Darth Panda, Darwinek, David Jordan, Decumanus, Discospinster, Drbreznjev, DubaiTerminator, Ed, Eeekster, Egmontaz, Everyking, Famartin, Fastily, FitzColinGerald, Foobaz, Gadget850, Garwig, Geologyguy, Gunnar Hendrich, Gurchzilla, Hamiltondaniel, Hammersoft, Happy5214, Hephaestos, Hibernian, Hike395, Hike796, Hmains, Hubertfarnsworth, Huw Powell, I dream of horses, Ikluft, JForget, JWB, Jackofifa10, Japanese Searobin, Jengod, Jojojigamobo, Jsayre64, Kbh3rd, Kjkolb, Klaus Bertow, Kmusser, Kopaka649, Kristen Eriksen, Kusunose, Kyleshimek, Lightmouse, LilHelpa, Look2See1, Looxix, Lpangelrob, Luckylettuce, Luna Santin, MONGO, MPF, Mackerm, Matt T2, Mav, MaxwellPerkins, Maylett, Mhaitham.shammaa, Mhockey, Michael Patrick, Mikenorton, Mild Bill Hiccup, Minna Sora no Shita, Mmcannis, MojaveNC, Motthoangwehuong, My76Strat, NE2, Neilarmius, NewEnglandYankee, Nivix, NorCalHistory, Northwesterner1, NuclearWarfare, Nuttycoconut, Optichan, Orlady, Peteforsyth, Pfly, Phaedriel, Pigman, Plastikspork, Plumpurple, Porqin, Prindleman, Prometheusg, Psemper, Qfl247, R'n'B, Ranjithsutari, RickK, Roundelmike, Rtdrury, SDC, Sardanaphalus, Serie, Shannon1, Skijackz, Spoon!, Stan Shebs, Staplegunther, Stepheng3, Sunray, Susfele, Tchoř, Template namespace initialisation script, Tenunda, Terryn3, Thatotherperson, The Thing That Should Not Be, ThinkBlue, Tide rolls, Tmangray, Toiyabe, Trevor MacInnis, Trimnik, Ufwuct, Ukexpat, Ulric1313, Valfontis, Vegaswikian, Vsmith, WBardwin, Wai Hong, Wars, Wavelength, Wbfergus, Wiki alf, Wilmabear2003, Wylandwombat, Zeimusu, Zibran 1, Zoicon5, 265 anonymous edits

Image Sources, Licenses and Contributors

Printed by Books on Demand GmbH, Norderstedt / Germany